Astronomers' Universe

For further vc
http://www.s]

David S. Stevenson

Extreme Explosions

Supernovae, Hypernovae, Magnetars,
and Other Unusual Cosmic Blasts

David S. Stevenson
Nottingham, UK

ISSN 1614-659X
ISBN 978-1-4614-8135-5   ISBN 978-1-4614-8136-2 (eBook)
DOI 10.1007/978-1-4614-8136-2
Springer New York Heidelberg Dordrecht London

Library of Congress Control Number: 2013945391

© Springer Science+Business Media, LLC 2014
This work is subject to copyright. All rights are reserved by the Publisher, whether the whole or part of the material is concerned, specifically the rights of translation, reprinting, reuse of illustrations, recitation, broadcasting, reproduction on microfilms or in any other physical way, and transmission or information storage and retrieval, electronic adaptation, computer software, or by similar or dissimilar methodology now known or hereafter developed. Exempted from this legal reservation are brief excerpts in connection with reviews or scholarly analysis or material supplied specifically for the purpose of being entered and executed on a computer system, for exclusive use by the purchaser of the work. Duplication of this publication or parts thereof is permitted only under the provisions of the Copyright Law of the Publisher's location, in its current version, and permission for use must always be obtained from Springer. Permissions for use may be obtained through RightsLink at the Copyright Clearance Center. Violations are liable to prosecution under the respective Copyright Law.
The use of general descriptive names, registered names, trademarks, service marks, etc. in this publication does not imply, even in the absence of a specific statement, that such names are exempt from the relevant protective laws and regulations and therefore free for general use.
While the advice and information in this book are believed to be true and accurate at the date of publication, neither the authors nor the editors nor the publisher can accept any legal responsibility for any errors or omissions that may be made. The publisher makes no warranty, express or implied, with respect to the material contained herein.

Printed on acid-free paper

Springer is part of Springer Science+Business Media (www.springer.com)

Preface

Although I was perhaps only 6 or 7 years old at the time, I remember my dad taking me out on cold winter evenings, pointing out the constellations as we walked slowly up the frost-glazed road. This was suburban Glasgow, then a city of one million people, and there was relatively less light pollution and many thousands of stars that were visible. I was immediately intrigued by the differences in brightness and colors of the stars, as much as I was about their twinkling effervescence. Even from a city as big as Glasgow, the sheer beauty of space never failed to amaze me.

I suppose it was my dad's enthusiasm for a night sky – a love born during his time on HMS Orion in the First World War – that really pushed me in this direction. By eight, I had developed an unhealthy love of explosions. Early on my main area of explosive interest, aside from James Bond movies, was volcanoes. However, once I knew that entire stars could blow themselves to pieces, I was never really going to look back. At eight, I had my first book on stars, and even though it was soon chewed to pieces by my pet hamster, it was still heavily thumbed in search of facts about why something as awesome as a star could blow up. By the time I was ten, I had already grasped how stars evolved, understood that there were different types of supernovae, and had absorbed some of the more peculiar aspects of the evolution of the universe.

In my teens, in the 1980s, I realized that there were very few books that extended what I already knew about stars, and although I was subsequently able to access peer-reviewed articles while at Glasgow and Cambridge Universities, it was evident that the number of more advanced titles were limited to the point of virtual non-existence.

The aim of this book is not to produce a comprehensive coverage of supernovae, nor is it going to rehash the general mechanisms of supernovae that can be found elsewhere in abundance. Instead, I wish to take this opportunity to present the more unusual examples of stellar deflagration and detonation that are pouring out of automated searches for supernovae. The science of stellar death has ballooned in the last decade, on the back of decades of theorization and observation. This is an attempt to bring those new understandings to a broader audience. In this title, I have attempted to cover as wide a range of discoveries as possible. However, if you trawl websites such as www.universetoday.com or www.ArXiV.com, you will undoubtedly discover more that I either could not include for reasons of space or continuity or simply was not aware of. However, this book should serve to extend considerably the baseline knowledge of those interested in stars and form a starting point for further personal discovery. I hope you enjoy it.

Nottingham, UK   David S. Stevenson

About the Author

David Sinclair Stevenson studied Molecular Biology at Glasgow University before completing his PhD in molecular genetics at the University of Cambridge. Subsequently, David studied and achieved a distinction in Astronomy and Planetary Science, and Geochemistry and Geophysics at the Open University. His peer-reviewed biological research articles from 1999 to 2009 include a paper on the early development of life, "The Origin of Translation," published in the *Journal of Theoretical Biology*.

David's interest in astronomy was encouraged from an early age by his father. This (combined with an interest in explosions!) has led David to research and write about the life and death of stars.

After a stint in academia, David became a teacher but continued to write scientific articles for various publications. He has published numerous articles on the Blackwell Plant Sciences website (2002–2007). "Turning Out the Lights" (an article about red dwarfs) was published in *Popular Astronomy* in 2003, "A Bigger Bang" (about Type Ia supernovae) in *Sky & Telescope* in July 2007, with "Supercharged Supernovae" featuring as the cover story for the October 2011 edition of *Sky & Telescope*. He is currently writing another book for Springer called *Under a Crimson Sun*, on the possibilities of life in red dwarf systems.

David lives in Nottingham in the UK with his wife and family.

Contents

Part I An Overview of Stellar Evolution

1. The Biology of Supernovae 3
    Introduction 3
    Spectra: Chemical Portraits of Stars 3
    Photometry: Behavioral Studies of Stellar Death 8
    The HR Diagram 10
    The Power Sources of Nuclear Reactions in Stars 15
    The Chemical Composition of Stars 18
    Stellar Structure 20
    Rotation and Angular Momentum 25
    The Effects of Magnetism 28
    The Biology of Supernovae 31
    Supernova Taxonomy 33
    Basic Identifying Features 35
    Conclusions 40

2. The Anatomy of Stellar Life and Death 41
    Initial Conditions 41
    The Life of a Star 45
    The Main Sequence 46
    β-Cephei Variables 52
    The Neutron Star 67
    The Fate of the Surrounding Star 70
    Formation of Supernova Remnants 73
    Messages From a Retreating Front 76
    Conclusions 79

Part II A Walk Across the Rooftops

3. Stellar Evolution at the Summit of the Main Sequence 83
    Introduction 83

Looking Deeper into the Controversy 84
A Problem with Wolves 87
The Humphrey-Davidson Limit 89
Luminous and Violent Blue Variables 91
Evolutionary Paths of the Most Massive Stars 95
Type IIn Supernovae 97
Explaining Type IIn Properties 99
Catching the Wave 100
From Imposter to Supernova 103
Direct Detonation of LBVs 105
Slow Blow: The Case of Supernova SN 2008iy 106
The Impact of Collisions (No Pun Intended!) 109
Conclusions 116

4. Collapsars, Hypernovae and Long Gamma Ray Bursts 119
Introduction 119
From Star Wars to Star Death 119
Beppo-SAX to the Rescue 120
Of Fireballs and Jets 123
A Best Fit: The Collapsar Model 126
Hypernovae and Hyperbolae 130
Reconstructing the Supernova-GRB Connection 136
Supernova: Or Supranova? 139
XRFs and Type Ibc Supernovae 142
SN 2010jp: The First Jet-Powered Type II Supernova 149
Conclusions 152

5. Death by Fallback 153
Introduction 153
The Mystery of Cygnus X-1 154
Controversial Supernovae 160
Populations 162
Conclusions 163

6. The Formation of Massive Stars by Collision and Their Fate 165
Introduction 165
A Lack of Interpersonal Skills: Harassment
and Scandal 166
Conclusions 173

7. **Electron-Capture Supernovae** 175
   Introduction 175
   Supernova or Imposter? 175
   The Troubling Fates of Intermediate Mass Stars 176
   Limiting Factors 178
   Post Main Sequence Evolution 180
   The Cassino da Urca 183
   Many Roads Lead to Rome 190
   Did the Progenitor of SN 2008S Spend
   Too Much Time at the Roulette Table? 191
   SN 2009md: A Faint Type IIP Supernova
   with a Troubling Origin 193
   A Coda from the Distant Past: Type I.5 Supernovae 196
   Conclusion 198

8. **Ultra-luminous Type IIn Supernovae** 199
   Introduction 199
   Taking the Pulse 199
   A Lethal Pulse: SN 2006gy and 2006tf
   as Pulsational Pair-Instability Supernovae 205
   SN 2008es: An Ultra-Luminous
   Type II-L Supernova 211
   A Deadly Couple Embrace: Was SN 2007bi
   the First Pair Instability Supernova? 217
   Was Pair Instability a Common Cause of Death
   in the Early Universe? 225
   Conclusion 227

9. **The Magnetar Model for Ultra-Luminous Supernovae** .. 229
   Introduction 229
   The Magnetar Model 230
   SN 2006aj and the X-ray Flashes 235
   Conclusions 236

10. **The Mysterious SN 2005ap and Luminous
    Blue Flashes** 239
    Introduction 239
    SN 2005ap 239
    Arrival of the White Knights 242
    The PPI Model 244
    The Magnetar Model 246
    The Buried GRB Model 247

Contents

Are These Blue Explosions Pair-Instability Events? 248
Pan-STARRS Events 248
Can We Tie Up All the Loose Ends? 249
Conclusions 250

Part III Thermonuclear Supernovae

11. Hypotheses and an Oxymoron 255
Introduction 255
What Astronomers Know 255
What Astronomers Think They Know 257
What Astronomers Don't Know 258
Mechanisms and Scenarios 258
The Single-Degenerate Scenario 259
More Problems 262
A Precious Few 263
Double-Degenerate Models 265
The Explosion 267
SN 2011fe: A Defining Type Ia Supernova 271
Conclusions 274

12. Are there Super-Chandrasekhar Supernovae? 275
Introduction 275
Type Ia Supernovae: A Reprise 275
SN 1991T 276
SN 2003fg: Too Bright for Its Own Good? 278
Loneliness in a Crowd 280
Conclusions 283

13. The Good, the Bad and the Ugly Thermonuclear Supernovae 285
Introduction 285
R Corona Borealis: Born Again Stars 286
Mergers of More Massive White Dwarf Stars 289
AM CVn Systems 293
Type .Ia Supernovae: Descendents
of AM CVn Binaries? 295
Record Breaking Type .Ia Supernova
or Something Else? 299
Do Some Type Ia Supernovae Explode
Without Detonation? 302
SN 2008ha: A Cousin of SN 2005hk? 305

PTF 09dav: Something Else? 310
SN 2005E: *Lex Parsimoniae*.. 313
Conclusions 318

Part IV In Flagrante Delicto

14. **The Mysterious Case of V838 Monocerotis
    and the Red Novae**... 323
    Introduction 323
    V838 Moncerotis 324
    V4332 Sagittarii 330
    A Stellar Merger Caught in the Act............................. 331
    M85's Red Nova... 335
    The Case of the Optical Transient NGC 300 OT............ 336
    Once Red, Now Blue 338
    What Do These Collisions Tell Us About
    Stellar Eruptions? 338
    The Great Eruption of Eta Carinae: A Reprise.............. 339
    Conclusions 342

15. **Between Scylla and Charybdis**................................. 343
    Introduction 343
    Type IIa? 343
    What Might the New Scheme Be Based Upon? 347
    Conclusions 350

Glossary 351

Index 365

Part I
An Overview of Stellar Evolution

1. The Biology of Supernovae

Introduction

Stars shredded into pieces and nearby planets boiling away are among the images conjured up by the word "supernova." Shockwaves pound outwards and turn rocky asteroids into cometary fluff, driving planetary shards from their orbits and showering neighboring star systems with waves of deadly radiation. It's dramatic stuff. But what really happens inside the blast wave of a supernova, and, more importantly how can we explain all the different forms of explosion that are emerging from automated supernova searches? This book explores these violent and sometimes strange landscapes, populated by the corpses of tortured stars. Recent observations have revealed a wealth of new types of explosion, as well as illuminating the underlying mechanisms of how stars explode.

A traveler of any sort needs a map, a compass and some sort of focal point onto which new discoveries can be pinned. Astronomers employ a number of tools that return the information they require to describe and explain supernovae. This chapter is subdivided into sections that build one upon the other. From these modest beginnings we will emerge in a world like no other, a world of poetic death, schadenfreude and humbling sacrifice.

We begin our journey with the tools to navigate: stellar spectra, the Hertzsprung-Russell diagram, photometry and a brief guide to stellar structure and function.

Spectra: Chemical Portraits of Stars

Spectroscopy can be thought of as the study of the "chemistry of light." When white light is passed through a prism or diffraction grating the light is split into its respective colors. Each chemical

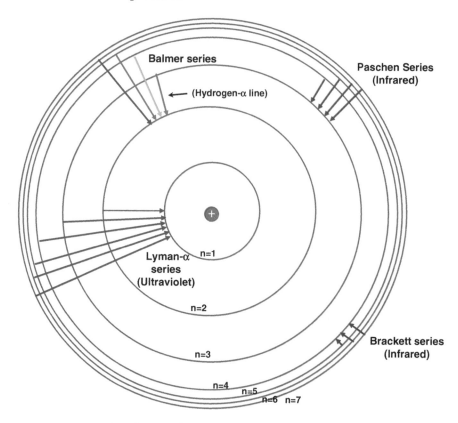

Fig. 1.1 Energy levels in a hydrogen atom. If electrons drop to the first level (n = 1) from the outer, higher energy levels they emit an ultraviolet photon. The well-known Balmer emission series (hydrogen-α) is produced when electrons drop from upper level three to the lower second level. Corresponding upward movements generate absorption features

element has a distinctive and unique spectroscopic signature. Although significantly lower in intensity than sunlight, starlight can also be split this way. The resulting spectrum gives a unique chemical signature for the star (Fig. 1.1).

Spectra come in three broad flavors: absorption, continuous and emission. A continuous spectrum is a bland rainbow of color produced by a chemically mixed source of light such as the surface, or photosphere, of a star. Where the light passes through intervening material an absorption spectrum of dark bands, called Fraunhofer lines, are observed. Finally, where light is emitted by excited gas this light is broken into bright bands called emission spectra.

Only the hottest stars or strongly irradiated gas produces emission lines. But what are these bands and how are they produced?

To answer this innocuous question we need to take a brief tour of the atom. In standard high school textbooks the atom is a Solar System in miniature, with planetary electrons orbiting a nuclear Sun. However, unlike planets whose orbits wobble and stretch, electrons experience an odd, restricted life. Electrons exist around the nucleus in specific energy levels or shells. Leaving aside the peculiarities of quantum physics, electrons remain in these shells with specific energies, unless they absorb or emit energy.

Now, not any energy will suffice. Electrons are picky. Thinking somewhat anthropomorphically, if the electron wants to move up one or more energy levels it must absorb a specific packet or quanta of energy – a photon with the correct frequency.

The energy levels are separated by specific energy amounts, and the absorbed photon must match the energy if the electron is to hop up. The electron can also move down a level by emitting a photon of light. This photon will exactly match the difference in the energy of the levels the electron has traversed.

Given enough energy, the electron can shake free of its bonds and escape the atomic nucleus altogether. This process, called ionization, allows the electron the freedom to absorb subsequent photons with any energy.

Absorption of photons by electrons gives rise to the dark Fraunhofer lines. Each element has a unique set of protons and (in the neutral state) an equal number of electrons that characterizes the chemistry of that atom. Consequently, the movement of electrons between atomic shells formulates a unique signature of absorption or emission that is evident in the spectrum of the light source or gas through which the light traverses.

An observer examining the light from the star, peering through the cloud, will see these absorptions as an absorption spectrum. Conversely an astronomer able to observe the gas without interference from light directly from the parent star will see only the light emitted by the same electrons as they lose energy and drop back down energy levels. This will produce an emission spectrum. In reality on Earth we can see all three of these from the same object (or at least the vicinity of the illuminating object) by blocking out portions of starlight entering the telescope's spectroscope. The end result can be a detailed chemical portrait of the star and its surroundings. Knowledge of this sort is essential, as we shall see that some stars have a deft knack of shedding their skin as they age. In these evolutions, the pupal case is often as important as the beast that emerges from underneath as they reveal the inner workings of the star in the period running up to its death.

Conversely, atoms with very excited, energetic electrons can calm down by electrons leaping down energy levels and emitting photons of light that correspond to the difference in energy levels. These liberated photons constitute the building blocks of the emission spectra – the bright lines on an otherwise dark background. Similarly, if light is emitted by a cloud of diffuse, hot gas it will also display emission lines. O-type stars, the hottest and most luminous in the universe, are sufficiently energetic to display emission lines corresponding to hydrogen and helium in their spectra. Any hydrogen or helium in their vicinity may also be ionized by the profuse emission of ultraviolet light from their surfaces. Consequently, the nebulae from which they form, and so frequently are associated with them, show prominent emission of hydrogen-alpha (Balmer) and Lyman-alpha lines and glow a profuse red. As a result O-type stars, buried within nebulosity and clusters of lesser stars, may be spotted over cosmological distances by the effect they have on nearby nebulosity. The Orion Nebula is perhaps the most famous nearby example powered by a small, tight cluster of O-type stars – the Trapezium. Many others are known (Fig. 1.2).

Consider this further. Imagine you are an electron and you're stuck on the ground floor of a building. You receive a package to deliver to the top floor. Now, you could take the stairs, but it's a long way to the top floor where the package needs posting. So, you take the elevator. The package is your photon, the particle of radiant energy that you collect to elevate you to the top floor. You absorb your photon, take your package and rise up the building in the elevator. In terms of the spectrum you could go all the way to the top if the photon is energetic enough, and a dark band will appear corresponding to this transition in the electron's energy. However, you could be left wanting and only have enough energy to reach the second floor. A different band, representing lower energy, will appear in the spectrum.

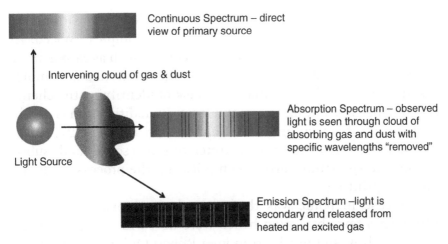

FIG. 1.2 The formation of different kinds of spectra. This is dependent on the viewing angle of the light source and the presence of any intervening material that is excited by short wavelength radiation or collisions

OK, let's say the photon packed enough punch to get you all the way to the top floor. At the top you take a rest, drop off your photon and drop all the way back down again. The photon emitted will exactly correspond to the one absorbed in terms of its energy. In which case, you might see an emission spectrum forming a mirror image of the absorption spectrum, but viewed at a different angle.

However, you fancy a change. On your way up you spotted some interesting things on different floors, so instead of going down straightaway, you pop out of the elevator onto different, intermediary hallways and landings before continuing your journey. Each time you drop off at an intermediate floor a photon of energy corresponding to the difference in energy levels is released. Since there are many different floors on the way down the chances are you will pop out at any number of these before returning to the lobby. Consequently, it is highly unlikely that the emission spectrum would directly mirror any absorption spectrum. There are simply too many possible other floors onto which an electron could emerge on its way back to the lobby. However, the emission spectrum would still carry the chemical fingerprint allowing the material's identification. The bands produced, whether absorption or emission, will still form the same unique chemical fingerprint for whatever chemical element is present.

In reality, unless the gas has a simple composition, the identification of all the bands may be very complex, indeed. As well as simple chemical elements, molecules, such as carbon monoxide or metal oxides in some stars, produce a myriad of complex absorption bands that can make the task of identifying the chemical composition vexing to say the least. In addition bands can be split in two if strong magnetic fields are present. Finally, stellar rotation and strong gravity can stretch or move some bands subtly around the spectrum, further complicating the process of chemical fingerprinting.

A case in point was the unusual supernovae SN 2005ap. Although a few absorption bands were noted in this supernova's spectrum, it wasn't until astronomer Robert Quimby realized that adjusting the spectrum for cosmological red shift meant that the explosion spectrum could be aligned with other, unusual blasts. At that point some of the chemical elements present were matched up to reference spectra and the problem was solved. When you look at the relatively simple spectrum of SN 2005ap you can appreciate how even a simple problem can cause confusion.

Photometry: Behavioral Studies of Stellar Death

Photometry is the measurement of the intensity, or flux, of energy in the form of electromagnetic radiation. The electromagnetic spectrum spans from radio waves at the lowest energy and frequencies through visible and ultraviolet all the way to the most energetic and highest frequency radiation – gamma rays. The majority of photometric examination of stars occurs from the near infrared through visible to the near ultraviolet. This is, in part, an historical situation, as stellar observations were most commonly done in or near to the visible portion of the electromagnetic spectrum. However, most stars emit the bulk of their radiation at visible wavelengths, hence the propensity of analyses at these wavelengths.

There are, however, many clear exceptions. The hottest stars (classes O and B) emit a large proportion of their energy in the ultraviolet range, while cool stars emit the bulk of their energy at infrared wavelengths. This means that measurements taken solely

Table 1.1 Pre-existing filter bands used in photometry and their corresponding wavelengths

Filter letter	Effective wavelength midpoint l_{eff} for standard filter	Description
Ultraviolet		
U	365 nm	"U" stands for ultraviolet
Visible		
B	445 nm	"B" stands for blue
V	551 nm	"V" stands for visual
G		"G" stands for green (visual)
R	658 nm	"R" stands for red
Infrared		
I	806 nm	"I" stands for infrared

in the visible portion of the electromagnetic spectrum will grossly underestimate the total energy (or flux) emitted by the object.

Nowadays, photometry is or can be done across the entire electromagnetic spectrum, but using a diverse set of instruments that are ground- and space-based. Past measurements of flux were initially executed utilizing specific photoelectric cells that measured the radiant energy striking each cell. However, with the advent of charge-coupled devices (CCD) multiple wavelengths can be examined simultaneously. Whereas in the past observations were made with single targets, CCD technology allows multiple targets to be imaged and analyzed simultaneously – vastly accelerating the acquisition of information. This has had an unprecedented effect on the rate of information gathering. One only has to look at the acceleration in the discovery of supernovae over the last decade. Various automated telescopic systems, employing CCD technology, now churn out hundreds of supernovae sightings per year. This has greatly expanded our pool of data and subsequently our understanding of the exotica present beyond the boundaries of our galaxy (Table 1.1).

The HR Diagram

Through the latter half of the nineteenth century and into the early part of this century sufficient numbers of stars were observed to begin the process of dissecting their morphology and gaining a better understanding of their underlying biology. Initial work by Father Angelo Secchi in the 1860s organized five broad groups of stars based on similarities and differences in their spectra. In the 1890s Henry Draper became the first person to photograph absorption spectra. Published in the Harvard Annals the work is part of the Henry Draper Memorial. Extending from this many astronomers will be familiar with stars bearing the prefix "HD," a derivation from his name. Work by Williamina Paton Fleming and Edward Charles Pickering fine-tuned and extended the classification scheme of Secchi, based on the spectroscopic analysis by Draper. Further work by Antonia Maury and later Annie Cannon refined the system further; Cannon's work eventually creating the classic O, B, A, F, G, K, M system that we know today.

It was the simplicity of this system of spectral classification that led Ejnar Hertzsprung (University of Leiden in the Netherlands) to realize that some of the stars were grouped according to luminosity. Some stars within one classification showed significantly lower "proper motion" (motion relative to the background stars) compared to others in the same class. This schism indicated that those with lower proper motion had to be further away from us than those stars showing the greater proper motion. Hence, these stars had to be significantly more luminous in order to appear equally bright.

Shortly thereafter, around 1911, Hertzsprung and separately in 1913, Henry Norris Russell (Princeton University), published subtly different but compatible diagrams. Hertzsprung's compared luminosity (absolute magnitude as the dependent variable) with color (Hertzsprung) and Russell's absolute magnitude with spectral type. However, as spectral type is a property of surface temperature, the two diagrams are effectively the same – hence the name Hertzsprung-Russell (or HR) diagram.

The eponymous chemistry tool, the Periodic Table, is an exquisite implement for dissecting and characterizing the chemical elements that make up our universe. It is such an easy tool to use, with elements grouped according to their properties, masses and atomic numbers. New elements can easily be slotted in, and clear patterns emerge linking elements in their groups. The underlying physics driving their chemical behaviors is also easy to determine with only a cursory examination of the Periodic Table.

Yet for all its finery, the Periodic Table still lacks the elegant simplicity of the HR diagram. Obviously it is worth recognizing that the 110 plus elements have a lot more inherent complexity compared to a star, but even so one could argue that the HR diagram is a more elegant tool than the Periodic Table. The virtue of this construct lies in its utter simplicity, its clear lack of obvious sophistication. Only two variables are present: color (or temperature), serving as the independent variable, and luminosity, functioning as the dependent. There is nothing else to this, and yet it is central to our understanding of all stars. The HR diagram is a blissfully simplistic representation that forms the backdrop for so much activity in astronomy.

The absolute magnitude scale is logarithmic, with each decrease in number corresponding to a jump in luminosity by a few times. Thus a star with an absolute magnitude of –1 is ten times brighter at the chosen waveband than a star with an absolute magnitude of 0.

Once constructed, the Hertzsprung-Russell diagram clearly demonstrated Hertzsprung's observed schism between dwarf and giant stars. They formed two distinct groups on the initial plots. The dwarfs formed a relatively smooth curve from top left (hot and luminous) to bottom right (cool and red), while the giants formed a separate locus positioned on the top right of the diagram (cool and red but luminous). There was a smudging of stars linking the two loci but above which few stars were present. This Hertzsprung gap is located in the region between spectral classes A5 and G0 and between +1 and –3 absolute magnitudes (i.e., between the top of the main sequence and the giants in a portion called the horizontal branch). It turns out that there is a fundamental reason for this gap: stars evolve so quickly from main sequence to red giant branch at this luminosity that they are rarely observed making this transition.

The study of stellar evolution tends to modify the HR diagram, converting the inferred parameter of temperature or spectral

12   Extreme Explosions

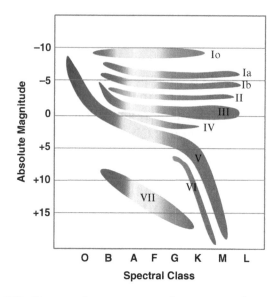

FIG. 1.3 The HR diagram in a more modern guise showing luminosity classes I through to VIII. *Io* hypergiants, *Ia* bright supergiants, *Ib* dimmer supergiants, *II* bright giants, *III* red giants, *IV* sub-giants, *V* main sequence, *VI* sub-dwarfs, *VII* white dwarfs. As well as inclusion of terms such as hypergiant and sub-dwarf, spectral classes L and T (showing lithium and methane respectively) have been added. These are primarily populated by brown dwarfs ("failed stars"), although a few bona fide hydrogen burning red dwarfs lie at the brighter end of class L – and the odd red giant. (Class T is omitted)

class into a photometric factor "B-V". This allows the flux of light at specific spectral wavelengths to be used as a reliable indicator of temperature, but one which is also numerical and hence easy to plot on a computer. These B-V versus absolute magnitude plots are referred to as color magnitude diagrams (Fig. 1.3).

Naturally, the diagram has evolved over time. During the 1940s the current system of luminosity classes O through to VII emerged. This MK luminosity system was completed by Morgan, Keenan and Kellman (hence the abbreviation "MK") in 1943 and is still in use today. It extended the original dwarf and giant groupings to six luminosity classes (see above). The giants have been subdivided into the following classes: "sub-giants – IV," "giants- III," "bright giants – II," "supergiants I" and lastly "hypergiants – O," each with increasing luminosity. All of the stars populating these regions of the HR diagram are more "evolved" and have left the portion of the diagram where stars fuse hydrogen into helium as a main energy source.

Dissecting the diagram from top left to bottom right is the main sequence – the region containing 90 % of the stars in the universe. This is also called class V. Main sequence stars fuse hydrogen to helium, a process that powers them for most of their lives. Running alongside the main sequence, from class G to M, is a parallel and more sparsely populated band of stars called the sub-dwarfs, or class VI. These are metal-poor stars that form an extension to the main sequence, comprising hydrogen-burning stars.

The supergiant locus (luminosity class I) has also been broken into two further subdivisions, Ia and Ib, with Ia forming the brighter clump. Descending numerically each class is brighter than the last, but with some overlap of the edges between these. Stars, after all, are not simple creatures, and like living things, the more you look at them the more blurred the boundaries become between classes. Consequently, a simple inspection of the HR diagram won't necessarily tell you what is happening to a star bearing particular characteristics.

An interesting biological example of a classification quandary in exists in parasites. They move around and behave like animal cells, and they lack the canonical cell walls of plant cells. Yet, if you look closely at the cells of the intracellular parasite plasmodium – the pathogenic cause of malaria – you will spot a cellular structure that is derived from a chloroplast. Chloroplasts are the site of photosynthesis, a classic plant trait. So is a plasmodium cell plant or animal?

A rather analogous example of this problem was revealed by the star Capella. Capella is actually two closely paired yellow giant stars and a more distant pair of red dwarf stars. The giant stars orbit one another so closely it took the power of spectroscopy to separate the partners in Capella's intricate dance. Capella's component stars were first resolved using interferometry in 1919 by John Anderson and Francis Pease at the Mount Wilson Observatory. This was the first interferometric measurement of any object outside the Solar System.

The Capella giants Aa and Ab have similar brightness and surface temperatures, which poses a problem for stellar evolution. How do two stars appear to evolve at the same time? After all, it is highly unlikely that they were born with identical masses. Even a slight initial difference would cause their paths to diverge as they age. Yet if we refer solely to the HR diagram as a guide to these stars, both appear to have reached the same evolutionary point at the same time.

The HR diagram offered few clues to this mystery. Both stars are rarities found at the red end of the HR gap and appear to have similar masses. Capella Aa weighs in at 2.7 solar masses, while Ab is slightly hotter and smaller, weighing in at 2.6 solar masses. Their luminosities are 79 and 78 times that of the Sun, respectively – uncannily close matches to one another. The question then repeats: Why are these stars so similar? Is this a simple fluke? The answer appears to lie in the internal state of each star. The slightly more massive Capella Aa appears to have completed its first ascent of the red giant branch and is now burning helium in its core. This has caused the star to contract, heat up and return to the yellow portion of the HR diagram. Ab, by contrast, with its lower mass, is lagging behind and is still on route from the main sequence to the giant branch. It is purely coincidental that both stars appear to be passing each other at the same location in the HR diagram. They are merely ships passing in the night.

A cursory inspection of the diagram didn't reveal this information. Instead the solution lay in the analysis of the chemical composition of the stars through spectroscopy. Star Aa has an envelope depleted in the element lithium. This element is readily destroyed in the hot interior of stars by nuclear fusion. However, the type of star Capella Aa came from would have probably resembled Sirius. As we shall see shortly, these stars cannot drag material from their outer layers down towards the core, where it is hot enough to fuse lithium. However, once these stars leave the main sequence and become red giants this process happens readily, destroying any lithium present. Capella Aa is depleted in lithium but Ab is not, suggesting that Aa has already become a red giant but Ab has not. The HR diagram thus offers clues to the evolutionary state of a star, but without additional information, certain mysteries will remain unsolved.

As a final coda to the HR diagram a little quirk must be emphasized. The orientation of the independent variable on the horizontal axis is inverted. Any high school boy or girl will tell you numbers increase along the axis, away from the origin. The HR diagram retains an idiosyncrasy of its inception based on spectral class. The highest temperatures are found closest to the origin – the point where the vertical axis intercepts it. In the color-magnitude diagram, this is corrected with numerical values increasing rightwards, away from the origin, but the location of the luminosity classes and their orientation is retained.

Quirks aside, the HR diagram remains a magnificent tool and central to modern astronomy teachings.

The Power Sources of Nuclear Reactions in Stars

Stars on the main sequence convert hydrogen into helium by the process of nuclear fusion. In outline, hydrogen combines in fours to make one helium atom. These reactions require high temperatures and high densities to flourish; therefore, they are generally limited to the stellar core, where temperatures are highest.

In the lowest mass stars nuclear fusion is a very sluggish process, and a star with one tenth the mass of the Sun will take nearly 6 trillion years to use up its fuel reserves. The Sun has greater reserves of fuel than a low mass star, but because it is more massive its core is hotter, and at higher temperatures nuclear reactions run through the fuel reserves a lot faster.

This is a general rule for all stars: the greater the mass the faster the fuel is used up. Stellar lifetimes extend from a little more than 10 trillion years for the lowest mass stars (approximately 0.075 solar masses) to less than 5 million years for the most massive (greater than 100 solar masses). Clearly this is a non-linear relationship, with three orders of magnitude in stellar mass equating to six orders of magnitude of stellar lifetime. The reason for this discrepancy is two-fold. For one thing, stars get hotter as their mass increases. Nuclear reactions are very sensitive to temperature, and only slight increases in temperature give very large increases in the rate of nuclear burning. This means massive stars burn fuel faster. Secondly, massive stars burn their hydrogen

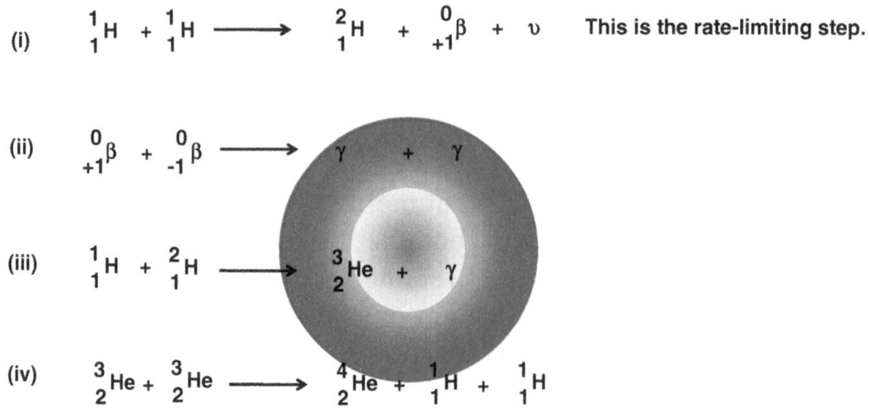

(i) $^1_1H + ^1_1H \longrightarrow\ ^2_1H + ^0_{+1}\beta + \upsilon$ **This is the rate-limiting step.**

(ii) $^0_{+1}\beta + ^0_{-1}\beta \longrightarrow \gamma + \gamma$

(iii) $^1_1H + ^2_1H \longrightarrow\ ^3_2He + \gamma$

(iv) $^3_2He + ^3_2He \longrightarrow\ ^4_2He + ^1_1H + ^1_1H$

Fig. 1.4 The Proton-Proton-I (PPI-I) chain. This series of reactions convert hydrogen into helium in four, sequential reactions. The first step limits these reactions to stars with more than 75 times the mass of Jupiter. The third step can occur in brown dwarfs with more than 13 Jupiter-masses, at temperatures above 700,000 K. β represents a positron, γ a gamma ray and υ a neutrino

fuel in a different manner from lower mass stars. The details will be described shortly, but the mechanism of hydrogen fusion in massive stars is more efficient and fuel is consumed more quickly than in low mass stars. These two effects contribute to a rapid decrease in stellar longevity as mass increases.

The Proton-Proton Chain of Low Mass Stars

There are three proton-proton chains in reality, but concentrating on that used most abundantly, the system can be outlined as follows. In the first step (which is also rate-limiting) hydrogen (a single proton) combines in pairs to form an isotope of hydrogen called deuterium, plus a particle of antimatter called a positron. The positron soon meets an electron and is annihilated, releasing two gamma rays (Fig. 1.4).

Deuterium, with one proton and one neutron, then combines at a much faster rate with another proton to form helium-3 plus a gamma ray. Lastly, pairs of helium-3 combine to form helium-4, plus two protons. The energy liberated is enormous and comes in the form of particle motion and the emitted gamma rays.

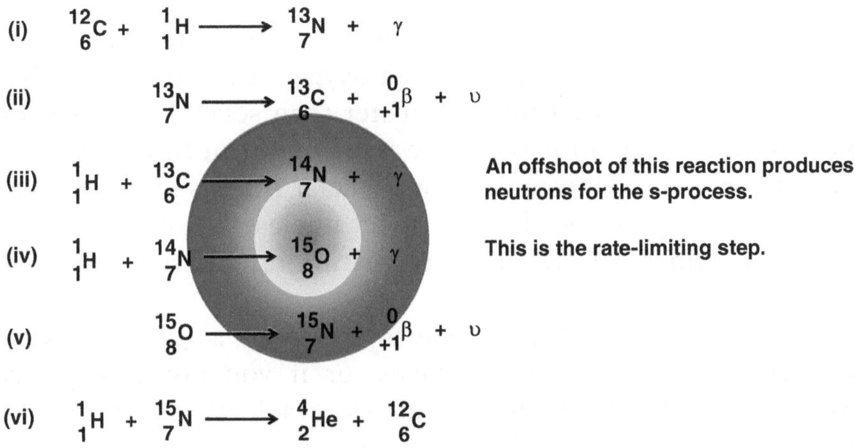

(i) $\ _{6}^{12}C + \ _{1}^{1}H \longrightarrow \ _{7}^{13}N + \gamma$

(ii) $\ _{7}^{13}N \longrightarrow \ _{6}^{13}C + \ _{+1}^{0}\beta + \upsilon$

(iii) $\ _{1}^{1}H + \ _{6}^{13}C \longrightarrow \ _{7}^{14}N + \gamma$ **An offshoot of this reaction produces neutrons for the s-process.**

(iv) $\ _{1}^{1}H + \ _{7}^{14}N \longrightarrow \ _{8}^{15}O + \gamma$ **This is the rate-limiting step.**

(v) $\ _{8}^{15}O \longrightarrow \ _{7}^{15}N + \ _{+1}^{0}\beta + \upsilon$

(vi) $\ _{1}^{1}H + \ _{7}^{15}N \longrightarrow \ _{2}^{4}He + \ _{6}^{12}C$

FIG 1.5 The carbon-nitrogen (CN) cycle consists of a series of reactions in which hydrogen is added sequentially to a carbon-12 seed nucleus. The CN cycle releases energy at a far higher rate than the proton-proton chain and is far more sensitive to temperature than these reactions. β represents a positron, γ a gamma ray and υ a neutrino. Some of the intermediates are unstable and release positrons. These reactions are useful on Earth as the source of positrons for PET (Positron Emission Tomography) scans. On white dwarf stars in binary systems that are burning hydrogen, annihilation of positrons with electrons releases copious heat. This leads to a runaway in the rate of nuclear reactions, which in turn leads to a nova explosion

The Carbon-Nitrogen Cycle

The majority of the stars dealt with in the second section of this book last only 3–10 million years at most, squandering their vast fuel reserves in the cosmological blink of an eye. The highly efficient carbon-nitrogen cycle powers this gluttony. Here, carbon-12, with 6 protons and 6 neutrons, fuses with a proton forming nitrogen-13. This decays, releasing a positron forming carbon-13. Carbon-13 fuses with another proton, forming nitrogen-14. These steps are fairly fast (Fig. 1.5).

Next, at a much slower pace, nitrogen-14 fuses with another proton, its third, forming an unstable oxygen-15 isotope, which also decays, releasing another positron. The product, nitrogen-15, finally fuses with a fourth proton, forming helium-4, while regenerating carbon-12 once more. Both positrons annihilate with electrons liberating energy in the form of gamma rays.

Once the stellar interior exhausts its supply of hydrogen, stars with masses exceeding half that of the Sun can go on to fuse their helium reserves to make carbon and oxygen. Even more massive stars, those with initial masses greater than seven times that of the Sun (the exact limit is a little uncertain) can fuse carbon to make neon and magnesium. Above 9 solar masses stars can create elements up to iron through direct nuclear fusion. The other chemical elements have more interesting origins that we shall return to later.

Nuclear fusion releases most of its energy as gamma rays – high energy electromagnetic waves, or if you prefer, photons (particles of light). Either term is an acceptable means of describing a gamma ray, because light likes to behave as particles and as waves.

The Chemical Composition of Stars

Stars vary significantly in their composition. *Significant* is an interesting word in science. It conjures up other words such as "large," "enormous" or "gargantuan." In reality, significant means different in a manner that can be distinguished via statistical analysis. For a star a difference of anything starting at 0.1 % can be significant (but is hardly spectacular numerically).

Variations in stellar composition only become extreme later in their lives as much of their initial stock of hydrogen is used up. Subtle variations in the initial makeup of the star can have profound effects on their subsequent evolution. This is especially true for massive stars – those with initial masses greater than 9 solar masses. Alterations in the amount of heavier elements can drastically affect how much mass the star holds on to as it ages, and this in turn affects how the star dies.

When a star is born its chemical composition, or chemotype, is set by the composition of the gas from which it forms. On a few occasions the star can be polluted by a close companion star, but leaving this aside it is the cloud of gas and dust from which the star is born that determines much of its subsequent fate.

The first stars born in the universe were made solely of hydrogen and helium, with a faint whiff of lithium, beryllium and boron.

After all, this was the inheritance these stars were bequeathed from the Big Bang. In terms of mass, this left a star with 76 % hydrogen and 24 % helium. The lithium comprised much less than 1 %, so we shall ignore it. However, subsequent stars were raised in nebulae polluted by the crematoria of these first stars. These stellar ashes gradually built up until by the point most of the stars in the Milky Way were formed, the Sun included, the ashes comprised 2 % of the total mass of the stars. Current stellar newborns inherit a larger mass of metals than the Sun (up to 6 % of their mass) and this can have profound effects on the ways these stars evolve as they age.

A general rule of thumb is stars with greater proportions of metals are larger, cooler and more luminous than their more impoverished siblings of equal initial mass. The presence of metals affects how readily energy can escape the stellar interior. You can think of a metal (any element heavier than helium) as a pot of electrons. These electrons will absorb photons of energy if they have the right frequency or wavelength. Metal atoms potentially have tens of electrons, all of which are able to absorb a photon with the correct characteristics.

Additionally, in the cooler climes of the stellar envelope, metals can form compounds with hydrogen or other metal atoms or ions. These compounds will also bear unique electronic signatures that allow them to absorb further frequencies of radiation. Therefore, metals have the ability to absorb much more of the incident radiation than hydrogen and helium alone. In turn this causes greater particle motion, which causes the star to inflate and cool.

Later in the life of the star, this more puffy structure can become more extreme. Capture of incident energy from the stellar core drives such vigorous particle motion that matter is driven readily away from the star – mass loss. The effect of metals on mass loss is a contentious one. It was assumed that mass loss in stars with a chemical composition like the Sun, but significantly more massive, would be extreme. Stars that began life with 100 times the mass of the Sun would become pared down by ferocious stellar winds long before they were able to explode as supernovae. This would limit the possible fates these stars could pursue. However, more recent measurements of mass loss indicate that it may be a lot lower than expected. There is much ongoing research in this area, but the effect of metallicity on stellar fate will undoubtedly produce more than the odd surprise as research unfolds.

Stellar Structure

Stars adopt structures that seek to provide greatest stability. A so-called main sequence star, one that is powered by the conversion of hydrogen to helium through nuclear reactions, has a relatively simple structure dictated by the force of gravity pushing inwards and the energy from nuclear reactions pushing outwards. Low mass stars are supported internally by gas pressure – the motion of particles within the star. The ultimate energy source propelling these particles is provided by the nuclear reactions occurring in the stellar core. Higher mass stars (those with more than around twice the mass of the Sun) are supported by radiation pressure as well as gas pressure. The relative contribution of radiation pressure steadily increases with stellar mass until it is effectively the sole source of support against gravity in the most massive stars.

The energy from the nuclear reactions has to make its way to the stellar surface, where it is radiated to space. Depending on the mass of the star, three broad structures are adopted by main sequence stars that bring about the efficient transfer of energy to outer space.

The Lowest Mass Stars (0.075–0.3 Solar Masses)

The lowest mass stars – those with less than 0.25–0.3 solar masses – behave like simple pans of boiling water. Nuclear reactions trickling away in their cores generate energy that is transported all the way to the stellar surface by convection. Their dense, treacly interiors are not conducive to the transport of energy by any other mechanism. Once the light reaches the stellar surface, the abrupt drop in density allows the light to wriggle free its bonds and escape into the near vacuum of space (Fig. 1.6).

Remember, the term "surface" is a little loose. There is no solid surface, merely a region where light can escape the gases. There is a drop in density and pressure, but in reality the photosphere

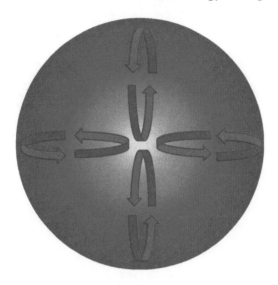

FIG. 1.6 The structure of a red dwarf star. Convection carries heat all the way from the nuclear furnace in the core to the photosphere, or visible surface of the star

is a narrow transition zone through which light escapes from the clutches of the atom and is released into space. As such the technical term is "photosphere" and applies to all surfaces where light is emitted, including the expanding debris from a supernova.

Low Mass Stars (0.3 to ~2.0 Solar Masses)

Somewhat more massive stars such as the Sun have a more complex structure. As internal temperatures rise and the overall density falls radiation is able to transport energy effectively. In the Sun this occurs in the inner three quarters of the stellar structure. Photons of light (gamma rays) released by nuclear reactions bounce between ions in the deep interior. Photons are successively absorbed and released, in a process that takes millions of years to transport them from their creation to the top of the radiative zone. Because photons take such a procrastinated route out of the deep interior, this has been called a random walk. Photons are absorbed by an electron changing its energy level, before the electron gives up the photon again in a random direction (Fig. 1.7).

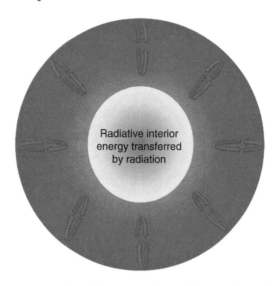

FIG. 1.7 The structure of a yellow dwarf star like the Sun. Convection carries heat only through the outer 25 % or so of the stellar structure to the photosphere. Radiation carries energy from the nuclear furnace through the remaining 75 % of the structure

It's a laborious process that props up the interior of stars such as the Sun against the force of gravity.

As the cooler outer portions of the star are approached, radiation becomes an ineffective means of transporting energy because of absorption of energy by ions and electrons, and through the processes leading to the ionization and recombination of hydrogen ions and electrons. The gas then becomes convective, roiling upwards in immense columns that carry the energy the remaining distance to the stellar surface.

Intermediate and Massive Stars (Masses Greater Than Approximately 2 Solar Masses)

There is a steady transition from stars that are fully convective to those that have convection limited to thinner and thinner portions of their outer layers. Radiation emerges as a transport mechanism in the stellar core of stars roughly one-third the mass of the Sun. As the mass increases to roughly twice that of the Sun (and the upper limit is a little contentious) the region of the star that

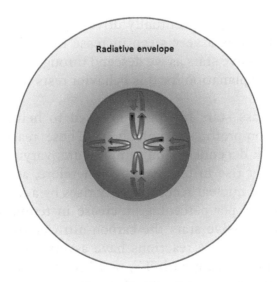

Fig. 1.8 The structure of a massive blue dwarf star like Spica A and B. Convection carries heat through-out the core. Radiation carries energy from just beyond the edge of the nuclear furnace through the remaining 75 % of the structure and out into space. Limited zones of convection, associated with the ionization of metals may drive convection in narrow bands within the envelope of the star

undergoes convection narrows upwards until radiation takes over the job of energy transport throughout the stellar envelope – that part outside the core. Recent work by Victoria Antoci has indicated that delta Scuti stars (1.5–2.5 solar masses) retain some form of convection just beneath their photospheres, indicating that there is quite a bit of overlap between stars with convective cores and those where convection is limited to the envelope. However, radiation is the principle means through which energy is transported through the envelope of massive stars – those with masses exceeding nine times that of the Sun (Fig. 1.8).

Finally, as we ascend toward the top of the main sequence, a few massive stars (δ Cephi stars) may have limited zones of convection within their otherwise radiative envelopes. They are driven by the ionization of metals. These massive stars are important because the pulsations allow astronomers to peer inside their interiors.

Meanwhile, in the stellar core, stars with more than 1.5 times the mass of the Sun begin to generate energy with such intensity that the energy output eventually drives convection in this part of the star. There is a very steep increase in the rate of nuclear reactions in massive stars as we descend through the core towards its heart. The explanation for this behavior rests with the carbon-nitrogen cycle.

In low mass stars hydrogen is fused to helium using the so-called proton-proton (or ppI)-chain. These reactions have a *relatively* weak dependence on temperature, varying with T^4, or four powers, per rise in temperature. Therefore, as temperatures rises, the rate of nuclear reactions only rises by a modest amount about four times per percentage increase in temperature. However, in more massive stars the carbon-nitrogen-oxygen (or CN) cycle dominates. These reactions have a very steep dependence on temperature, varying with T^{17}, or 17 powers. A 1 % increase in temperature means the rate of reaction increases by nearly 20 times. Put another way, if you double the temperature, the proton-proton cycle goes about 10–20 times faster, but the CNO cycles goes billions or hundreds of billions of times faster.

So, as we dive deeper into the heart of a massive star, the rate of nuclear reactions accelerates rapidly, which in turn accelerates the amount of energy released. The increase in rate is such that only convection can efficiently transport the energy produced. The radiation is simply not sufficient. Although the CN cycle takes over the main job of hydrogen fusion in stars, with initial masses exceeding 1.5 solar masses, it isn't until the star is 2–2.25 solar masses that sufficient energy is created to drive convection throughout the bulk of the core. At lower stellar masses convection is confined to the innermost core, but this zone steadily increases in size until the whole core is broiling.

Massive stars are, therefore, dominated by energy transfer via convection in their deep interiors and by radiation in their envelopes – the portion of the star surrounding the core. Therefore, with the exception of a thin sliver of the Hertzsprung-Russell diagram, main sequence stars either convect in their outer layers or in their cores but probably not both.

Importantly, models suggest that convection in the core may overshoot somewhat, so material from the base of the radiative envelope can get mixed into the roiling core. Overshooting may contribute to the enrichment of the lower part of the envelope in the products of hydrogen fusion – isotopes of nitrogen in particular – as well as bringing more fuel into the core. These nitrogen isotopes are produced by the carbon-nitrogen cycle and may become visible in the spectrum of some massive stars as they evolve away from the main sequence (Chap. 3).

Rotation and Angular Momentum

Momentum is a property of moving matter. Momentum is simply a product of the mass of a moving object and its velocity. A moving truck or person has momentum, and like energy it is always conserved. That means when a moving object hits another, the momentum of the original moving object is shared out between the two bodies following the collision. Think of a fired gun. Before the trigger was pulled, the gun and the bullet weren't moving and their total momentum was zero. When the trigger is pulled and the bullet launches forward it has forward momentum. The gun then recoils. Its momentum must equal that of the bullet, but have the opposite sign as its moving in the opposite direction. The momentum at the end is equal to that at the start.

All stars are born rotating. Even slight differences in the motion of the gas and dust within the parent nebula confer rotation to the nascent star. Consequently, the mass in a star has *angular* momentum. Angular momentum is a property of how fast the star (or part of the star) is rotating on its axis multiplied by the mass of the rotating material. It varies with distance from the axis of rotation, being greater per unit mass with increasing distance.

When a rotating body shrinks and the mass moves closer to its center, angular momentum is conserved and the body must rotate faster. Therefore, as a star forms from its parent nebula the momentum of the material in the nebula is conserved within the developing star. However, if you work out how much angular momentum a star would have at birth it vastly exceeds what is observed. Stars must shed some momentum during their formation.

Conversely, if the body expands mass moves further away from the center and the speed of rotation must go down or angular momentum will not be conserved. You can see this at the ice rink.

An ice skater brings his or her arms inwards to spin faster, while doing the opposite to slow down.

Think about a merry-go-round. When a child leaps onto and moves to the center of the roundabout, it speeds up. If he or she moves to the outside it will make it slow down.

In stars angular momentum comes into play in many different settings that are important in later evolution. As the star is condensing out of its parent nebula, the nascent star contracts, and the rate of spin increases. Similarly, at the close of the main sequence the helium-enriched core contracts. This causes its rate of rotation to pick up. Meanwhile around the core, the hydrogen-rich envelope expands, and the rate of its rotation slows down. The difference in rotation rates of each part of the star can lead to some mixing between the layers. In principle the increase in the angular momentum of the core is balanced by a reduction in angular momentum of the outer layers. Loss of material from the envelope in stellar winds removes some angular momentum, causing the envelope to further slow its rotation down.

In close binary star systems angular momentum is shared between the two stars. If one star begins shedding material there can be some complex exchanges of angular momentum between the two objects and surrounding space. This exchange of momentum can lead to some exciting interactions between the two stars, resulting in fireworks. In the final section some of these scenarios are examined to illustrate how angular momentum influences the course of stellar death.

Later on we will also examine the potential role the conservation of angular momentum has in the decimation of the universe's most massive stars. Most models of massive Population III (primordial) stars indicate that many would be so massive that they would explode via the so-called pair-instability (PI) route (Chaps. 3 and 8). Without going into detail, supernovae derived from stars with masses in excess of 130 times that of the Sun are expected to produce very large quantities of iron group elements (iron, cobalt and nickel). However, observations of Population II stars, the oldest stars in the galaxy and the immediate descendants of the Population III stars, show little evidence of this kind of chemical enrichment. Instead most show enhancements in the abundance of intermediate mass elements – the kind produced by conventional core collapse supernovae. Therefore, on the one hand, we have models suggesting particularly massive Population III stars, but on the other we see no evidence these stars were especially substantial. How can we square this circle?

The solution may rest with angular momentum. Modeling by André Maeder (Geneva Observatory, Switzerland) and others, published in the journal *Nature* in 2011, suggests that the earliest stars may have been fast rotators or, as Maeder coined, "spinstars." These rapidly rotating stars would have had sufficient reserves of angular momentum to throw material off their equatorial planes. A combination of fast rotation and high surface radiation would effectively shed material from the stellar surfaces, helping reduce the mass of the stars throughout their hydrogen-burning phase. This reduction may have proved sufficiently extreme that these early stars would have avoided pair instability and died more sedately as core-collapse supernovae. Certainly, this would help solve a puzzle.

A final area worth mentioning here is the recent series of observations that some massive stars called luminous blue variables (LBVs) sometimes explode as supernovae. Early work on massive stars indicated that most LBVs would shed their hydrogen-rich envelopes through stellar winds and more violent pulses, leaving a more slender Wolf-Rayet (WR) star behind. It was these WR stars that went on to explode as supernovae. The obvious question is, why do some LBVs explode before they become WR stars? The suggestion is, and these are still very early days, that those massive stars that are fast rotators become WR stars early on, shedding their hydrogen skins through the same two effects described for spinstars – fast rotation and profuse stellar winds. Slower rotating massive stars have envelopes with lower overall kinetic energy and are, therefore, more able to hold onto them as they evolve towards their ultimate demise.

Angular momentum is also fundamentally important during stellar birth. The imprint of this physical property is seen in powerful jets shed by infant stars. As the infant star contracts excess angular momentum is shed through winds and jets that launch from the proto-stellar poles. And as it is at birth, it is at death. There is ample evidence for the role of angular momentum in the launching of jets by white dwarf stars, neutron stars and black holes.

As these remnants form and contract, excess angular momentum is shed through jets launched from the rotation or magnetic poles of these corpses. The conservation of angular momentum, this fundamental and inherent property of all rotating objects, thus gives rise to some of the most spectacular objects in the universe.

The Effects of Magnetism

All stars are born with some inherent magnetic field. As stars age, unless they generate a field of their own, these fields should weaken. Thus over the time the star spends on the main sequence magnetic field strength declines and can be used as an indicator of stellar youthfulness.

The Sun, for example, has an average field strength of around 1 gauss (1 G), or twice as strong as that of Earth. Sunspots have a field up to 2,000 times stronger, but this is localized in its reach. One gauss is broadly comparable to a refrigerator magnet. Main sequence stars may have significantly stronger fields, but the real beasts of magnetism are the magnetars, with fields over 100 trillion times stronger than the Sun's.

Where magnetic fields are particularly strong, they have an interesting effect on stellar spectra. The so-called Zeeman effect is the splitting in two of spectral absorption bands by magnetic fields. Although this effect is not readily apparent in main sequence massive stars it is visible in their progeny, the neutron stars formed through core collapse. Before we look at these, we need to consider where the stellar magnetic field arises.

Throughout interstellar and intergalactic space there is a very weak magnetic field permeating the vacuum of space. The interstellar field is incredibly weak by terrestrial standards (around 0.00001 G), but over interstellar distances it can still have profound effects. It modulates the movements of cosmic rays as they stream away from supernova remnants and it also has profound effects on the formation of stars. Interstellar fields have the propensity of holding up clouds of gas against collapse under gravity, and by guiding cosmic rays interstellar fields affect the ionization state of interstellar hydrogen and the formation of exotic molecules from diatomic hydrogen (H_2) through water and even alcohol and amino acids. This is no trivial thing. These molecules help cool the clouds of hydrogen gas and ultimately breathe life into the cosmos by triggering star formation.

When enough hydrogen and helium has accumulated, gravity takes hold and begins weaving the building blocks of the star in and out of the interstellar magnetic field and downwards and inwards, triggering the formation of a protostar. (There's more on the process of star formation in the next chapter.) During the collapse of the protostar, convection is the primary mechanism through which energy is transferred from the deep interior to space. Protostars abundantly display evidence of magnetic fields through sculpted and collimated outflows guided by these fields. However, once a star hits the main sequence, many stars have limited or negligible convection, and the question arises: can they retain the magnetic fields they had at birth?

Stars that have convecting outer envelopes (all stars less than about 1.5 solar masses) generate a magnetic field internally through a combination of convection and rotation. For example the Sun's relatively modest field is generated near the base of the convection zone. The precise mechanism is not yet fully understood.

Massive stars prove to be a little more difficult to figure out. A number of massive stars show some evidence of magnetic fields. In general this property is often determined indirectly through measurements of the X-ray flux from these stars. A few massive stars show periodicity in the strength of X-rays generated in the stellar wind flowing from the star. This variation appears to match the stellar rotation rate. The suggestion is that the magnetic field of the stars is not directly aligned with the rotation pole of the star – just as is the case on Earth. Therefore, as the star rotates the stellar wind alters in strength, depending on the direction the magnetic field is pointing.

Theta Orions C is one such star. Theta Orions C (θ^1-Orions C) is the most massive Trapezium star in the Orion Nebula (M 42), weighing in at approximately 40 solar masses, with a spectral class of O7V. Theta Orions is both very young (at most 1 million years) and the principle source of ionizing UV radiation in the Orion Nebula, but it also displays variability in its X-ray, Hα and He I emission. The X-ray variability matches the variability in the Hα emission with a period of 15.422 ± 0.002 days. In a manner closely analogous to less massive Be stars, the variability in X-ray and other emissions is best explained by a rotating disc embedded in a stellar magnetic field that is inclined to the line of sight. Jacques Babel and co-worker Thierry Montmerle suggest that θ^1-Orions C has a magnetic field of the order of ~300 G, or 300 times stronger than that of the Sun. The presence or absence of magnetic fields in massive stars is not a trivial issue, as it could have profound effects on the way these stars lose mass and on how they evolve further to produce neutron stars or black holes.

The question then arises, how do massive stars generate this field? In principle a magnetic field could be generated in the spinning, convecting core while they are on the main sequence. Alternatively, they could hang onto the one they were born with, until they leave the main sequence. A third model suggests that the field is generated in a thick circumstellar disc that orbits in the equatorial plain of the star. However, it has proven difficult to model the transport of the magnetic field through a star's relatively static radiative envelopes or to demonstrate how a massive star could support the disc around its girth.

Work by Jacque Richer and colleagues in Montreal investigated the structure of the envelopes of intermediate mass stars (stars with roughly 2–9 solar masses). They showed that Population I, intermediate mass stars could generate convection in their normally radiative envelopes if a compositional gradient is developed. In the radiative envelope iron would settle downwards into a layer some way below the surface of the star. Here the enhanced abundance of metals would increase the absorption of radiation, steepening the thermal gradient and thus generating a narrow zone of convection in this portion of the envelope. Indeed, Richer's models show that 2.5 solar mass stars have, initially at least, three convective layers. The iron-convection zone is the deepest and found where the temperature is ~200,000 K. Atop this are two narrow helium convective layers (the lower of these is the layer that drives pulsations in F-class stars at 50,000 K) where He I and He II ionization occur. However, convection in these layers disappears very early on as much of the helium drains downwards into the core of the star. Without the necessary abundance of helium, there is insufficient opacity to drive convection.

There is no problem with generating magnetic fields in the cores of massive stars in which core convection is occurring. The fluid motion of the ionized (and strongly conducting) gases is more than sufficient to generate a strong magnetic field. Paul Charbonneau, Keith MacGregor (then at the High Altitude Observatory, National Center for Atmospheric Research, Boulder) and Joseph Cassinelli (University of Madison) carried out extensive work in this area in the late 1990s and early 2000s. There was a problem with getting the field to the surface of the star, but in principle the star could generate and manifest its own external magnetic field. At present, although some scientists have made suggestions as to how this might occur, the generation and transport of magnetic fields in massive stars still seems problematic.

We will now push on, away from the basics of stars and into the "biology" of supernovae. Beginning with the basic classification of these explosions, we will then further dissect the different types of explosion, and begin our journey towards understanding the most unusual explosions in the universe.

The Biology of Supernovae

Now, you might think that biology and supernovae have very little in common, and superficially you'd be correct. However, on closer examination the sage words of the English-born philosopher, George Henry Lewes – an influence on Charles Darwin – applies: "Science is the systematic classification of experience." Indeed one is usually more familiar with the classification of organisms than with stars. Yet, when you examine the field of astrophysics, you soon discover that it is rife with biological concepts and phraseology. Astronomers have systematically classified the stellar zoo in a way that would make any self-respecting biologist proud.

In biological circles classification is often the first of many steps towards an understanding of an underlying principle. Evolutionary science is a prime example. Beginning in the eighteenth century and continuing throughout the nineteenth and twentieth centuries biologists have sought to identify organisms and classify them according to their characteristics. The principle is simple. If the organisms share common features they may behave

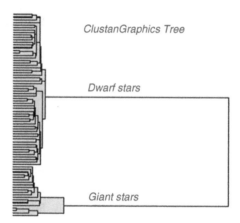

FIG. 1.9 A clustan graphics analysis of a selection of stars. The analysis clearly separates the stars into giants and dwarfs based on their properties. Such analysis could be used to distinguish and classify supernovae (Source: Credit: David Wishart, University of St. Andrews. Used with permission.) (http://www.clustan.com/clustering_v_decision_trees.html)

in biologically similar ways; be related in evolutionary terms; and finally, an understanding of one organism may henceforth shed light on the origin and behavior of another. This principle of taxonomy underlies several areas of biology and was first standardized by Carl Linnaeus, coining the term Linnaean classification.

Similarly, in astrophysics, classification of stars led first to the creation of the eponymous and essential tool – the Hertzsprung-Russell diagram – and later to an understanding of how stars work. In biology, taxonomy led to phylogeny: the groupings of organisms in evolutionary terms. The study of DNA later revolutionized this field by identifying the specific underlying features, the genes, that relate organisms. In astrophysical terms phylogeny can be understood in terms of the underlying features of stars, such as the types of nuclear reactions that occur inside stars, the internal structure of stars and the external behavior of stars, such as the loss of mass through stellar winds (Fig. 1.9).

Taking this one step further, if you visit the "Clustan" website (http://www.clustan.com/clustering_v_decision_trees.html) you can see a phylogenetic approach to the ubiquitous HR diagram (see above). Stars were clustered according to luminosity and temperature, and the output demarcated the giant and dwarf stars as well as subdividing different stars within these groups according to their stellar phenotypes, or appearance. It seems there are no bounds to the links between biology and stellar astrophysics.

Indeed we can take a slightly more whimsical approach that might make a geneticist proud. In genetics we consider the genotype and its impact on the phenotype. To a geneticist the genotype is the genetic content of the organism – its genes. The phenotype is the appearance or behavioral properties of the organism. We could easily extend this nomenclature, in modified form, to stars. Instead of genotype we might consider "chemotype" – the precise chemical composition of the star. Extending outwards from this core property, the phenotype would represent an identical property to that in biology: the appearance and behavior of the star. The link between chemotype and phenotype is identical to that between genotype and phenotype. Yes, whimsical perhaps, but whereas the commonly referred term "stellar evolution" would to a biologist seem something of a misnomer, chemotype and phenotype would apply precisely.

Why is stellar evolution a misnomer? Well, evolution refers to the change in complexity of a species over time, not an individual organism as it grows older. To a biologist the changes in an individual over time are called aging, not evolution. You don't evolve as you age. At least in an etymological sense chemotype and phenotype are precise. Therefore, they are worth pursuing.

Whimsy aside, as we look more closely at supernovae, again taxonomic classification underlies our growing understanding of the processes involved in stellar death.

Supernova Taxonomy

The term supernova (originally "super-nova") was coined by Fritz Zwicky during a lecture he gave at Caltech in 1931. Observations by Zwicky and Walter Baade during the 1930s had revealed what they initially believed were a subclass of the well known and more frequent explosions called novae. These *super*-novae were clearly a lot more powerful than the novae, but without a mechanistic explanation to discriminate them from each other, they were initially bunched together. In 1938, after the term had seeped into the modern astrophysical dialect, the hyphen was dropped and the currently used term emerged.

The origin of the current classification system goes back to the 1940s. Based on 14 supernovae, Minkowski subdivided the supernovae he saw into a relatively homogeneous population of hydrogen free events he called Type I. The others, a very heterogeneous group, were referred to as Type II. They showed hydrogen in their spectra. Later Fritz Zwicky proposed adding classes III, IV and V. These were all hydrogen-rich events. His class V events are now associated with so-called supernovae-imposters – weak Type II events that we shall return to in Chap. 6.

The spectral division of Type I supernovae and Type II events can broadly be compared to the division of plants and animals as separate species. OK, hardly an inspiring leap compared to what we appreciate now, but it was certainly a first big step along the road to understanding these explosions. As is the case with biology, further observations of living creatures allows additional features to be identified. We don't just have cats; we have lions, jaguars, tigers, lynx, etc. Similarly, Type I supernovae are now subdivided by spectra into Type Ia, Type Ib and Type Ic with a very recent and slightly scurrilous addition "Type .Ia." More on that one later.

Finally, during the 1990s, the Type II supernovae were grouped according to features in their light curves. The majority of Type II explosions have a plateau phase during the decline in their light output following the explosion. This gives them the suffix – "P" (Type II-P). A minority of Type II explosions show different characteristics in either their light curves or in early spectra that set them apart. You can think of the stripes in the stellar spectra as mere projections of the stripes and markings of wild animals. Stellar astrophysicists are the zoologists seeking to understand the wild animals that roam the skies.

This chapter will identify and dissect the various types of supernovae from core-collapse to thermonuclear and lay the groundwork for subsequent investigations. We will then take an overview of the processes that give rise to different types of supernovae.

Basic Identifying Features

Supernovae can be subdivided first of all on the presence or absence of hydrogen in their spectrum, as was described above. This is analogous to separating organisms into living or extinct and not much more. Closer scrutiny of each beast reveals a plethora of other details that allow finer dissection of each supernova. Like an anatomist disseminating the viscera of an animal, an astrophysicist dissects the light of the blast (Fig. 1.10).

The following is an outline of the spectral features found in each type of supernova. It should become obvious as we advance through this book that further subdivisions and genera emerge that are eventually assigned their own classification or merged with pre-existing classes. By the time this book goes to press several more species of supernova might have emerged.

Type Ia

These are hydrogen-deficient supernovae that display abundant iron group elements in their spectra as well as a defining

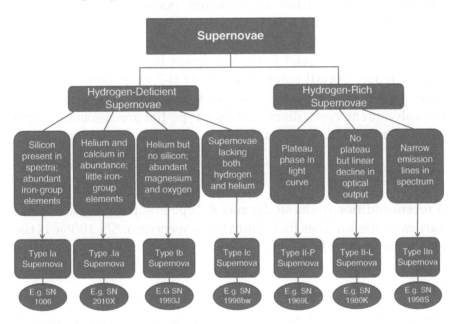

FIG. 1.10 A supernovae classification scheme, based on underlying spectral properties and light curves

absorption band for silicon at 615.5 nm (nanometers, or billionths of a meter). The silicon and iron group elements are the most abundant products of nuclear reactions in these explosions, hence their prominence in the spectra. Late spectra are dominated by iron group elements synthesized in abundance in the explosion. These supernovae are best explained by the destruction of carbon- and oxygen-rich white dwarf stars.

Type .Ia ("Point One-a")

These are faint Type Ia-like events, roughly a tenth as bright as their showier cousins (hence the somewhat frivolous name). They have fast-evolving light curves and show some differences in the types and proportions of chemical elements that are synthesized. SN 2010X is a well-studied example. Again, these are almost certainly caused by some form of detonation on or within a white dwarf star.

Type Ib

Prominent helium lines at 587.6 nm but no silicon. Oxygen, magnesium and other intermediate mass elements, such as calcium, are also usually prominent. Light curves of Type Ib (and Type Ic) explosions tend to peak at lower luminosities, and their light is redder overall compared to Type II-P explosions. However, there are exceptions to this rule, as we shall see. Type Ib and all the subsequent types discussed (bar Type IIa) are associated with core collapse in massive stars.

Type Ibn

A recent addition to the stellar zoo. A Type Ib supernova that shows narrow emission lines for helium in its spectrum. SN 2006jc is the prototype of this class.

Type Ic

These explosions lack hydrogen or helium in their spectra and fall into two distinct classes, broad- and narrow-lined. The broad-lined explosions have been associated with long gamma ray bursts; it is the high (relativistic) velocities of their ejecta that generate the broad lines. SN 1998bw is a prototypical member of this class of explosion. Narrow-line Type Ic explosions are more common and show narrower absorption lines in the spectra.

Type II-P

In many ways these are the quintessential core-collapse supernovae. These explosions mark the death of massive stars that have retained a large hydrogen-rich envelope. The spectra of these supernovae reveal so-called P-Cygni lines. P-Cygni lines are generated by a shell or blobs of gas moving outward and away from a central light source – in this case a supernova – but it could be a central hot star in other situations.

The outflow of gas that is not projected against the central source produces an emission line, as we have already seen. This line may be broadened by the rapid expansion of the gas away from the central illuminating source. This is usually not Doppler-shifted by the expansion, as most of the movement is perpendicular to the line of sight. Gas in front of the explosion that is moving towards the observer will produce blue-shifted absorption lines. The absorption line will appear at the blue edge of the emission line, giving a characteristic profile. Hydrogen is obviously prominent at early times in the spectra, but oxygen and other intermediate mass elements become evident later as the hydrogen-rich envelope disperses.

A key feature of these supernovae is a plateau in the light curve. The size of this can be very variable but demarcates a stage in the cooling of the expanding supernova remnant as hydrogen ions capture electrons and release energy as light. This process of recombination keeps the supernova shining at near constant brightness until the photosphere of the explosion has descended far enough into the expanding debris that no hydrogen remains.

Type II-L

Hydrogen Balmer lines are again present, but there is no plateau in the light curve. This may be due to the progenitor star having a very thin hydrogen envelope prior to the explosion, or possibly the pollution of the light curve by a fast-moving jet that punched through the star towards the observer, generating additional luminosity at the peak.

Type IIn

One of the newer classes of hydrogen-rich supernovae, Type IIn explosions have recently garnered a lot of attention. These display narrow emission lines instead of classic P-Cygni absorption features in their spectra. The emission lines are generated as the flash from the outgoing shock irradiates and ionizes hydrogen-rich gas around the explosion. Additional intermediate and broad emission lines appear to demarcate the point where the supernova blast wave impacted hydrogen-rich gas surrounding the doomed star.

Type IIa

This is a somewhat contradictory class of events, at least in terms of nomenclature. These explosions appear initially as classic Type Ia events with no hydrogen in their spectra. However, at later times, prominent lines of hydrogen appear, indicating the blast wave of the Type Ia explosion has impacted a shell or disc of hydrogen-rich gas lying further away from the explosion. The hydrogen may have come from a companion star or from some form of common envelope that surrounded the moribund star and its companion (Figs. 1.11 and 1.12).

Type IIa supernovae lack any obvious link to Type IIb events such as SN 1993 J, as their names might suggest. Given the lack of any relationship between these types of explosion one is forced to concede that the classification system may be ripe for an overhaul. Indeed, the development of the current system has been a bit like taking a 1920s Ford and subtly and continuously adding new features to it, then seeing if it still runs. The still evolving taxonomic scheme employed by astronomers seems increasingly to be like tinkering. Perhaps a more brutal revamp might be in order – particularly as completely novel forms of explosion are emerging, and the mechanisms that drive these are becoming clearer as well.

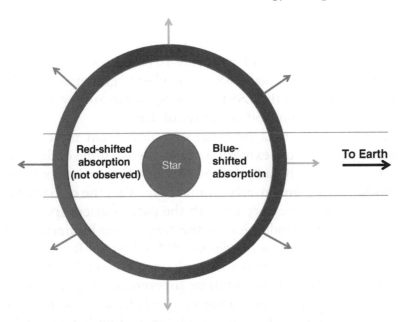

FIG. 1.11 **The creation of P-Cygni lines.** Material moving toward earth produces blue-shifted absorption features, whereas material not backlit by the central source produces an emission line

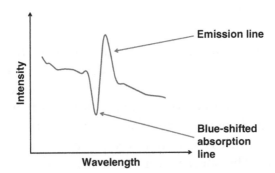

FIG. 1.12 Classic P-Cygni profile (paired emission and absorption – usually of hydrogen) caused by material expanding away from a central illuminating source

To illustrate the problem, consider the growth in data. From the 1950s to the 1990s (i.e., the best part of four decades) the Digitized Sky Survey (DSS) churned out 5 terabytes of data. To put this in perspective, a decent-sized PC or Mac computer will have a hard drive with similar storage proportions. By 2010 the Sloan Digital Sky Survey (now in its third incarnation) had amassed ten times this amount of data. This was after a decade of bagging. By contrast, the prospective Pan-STARRS 4 (Panoramic Survey Telescope and Rapid Response Finder) will produce 3–10 petabytes – or 1,000 times the output of the original DSS. Somewhat frighteningly, by 2018 (budgets permitting) the LSST (Large Synoptic Survey Telescope) may amass a whopping 50 petabytes of data.

Not all of this data concerns supernovae, but there is a vast potential to find these explosions in the dense mine of data. Such telescopic surveys broadly follow Moore's Law – the doubling of computing power on a nearly annual basis. Can the old supernova classification scheme keep up with the pace of discovery?

The recent introduction of the term super-luminous supernovae (SLSN), with its subdivisions, SLSN-I, SLSN-II and SLSN-R, is even less helpful. This is a jet pack on a blow-up balloon. What are the relationships between these supernovae? Are SLSN utterly distinct from the previously characterized classes or do they share mechanisms that would be better understood with a less divisive name?

Conclusions

As we step boldly through the second decade of the twenty-first century the number and types of supernovae astronomers are discovering seems to be increasing without bound. At present, although we understand rather a lot about how supernovae happen and the nature of some of their forms, the wealth of data streaming in from countless automated supernova searches is almost overwhelming. The sheer volume now demands a better scrutiny of the blast mechanisms, and this in turn requires a revolution in the systems used to classify these events. SLSN are a case in point. Although not quite coming from nowhere, these rare yet powerful explosions seem to demand their own system of classification. However, this is a distraction. A classification system stands or falls on its ability to incorporate the shock of the new. At present Zwicky and Baade's 1940s system is showing more than the odd crack and needs thorough reform. Tackled appropriately something magical may emerge that not only classifies explosions but leads to a better understanding of how they occur.

2. The Anatomy of Stellar Life and Death

Stars begin as clouds of gas and dust, called nebulae, with temperatures hovering a few degrees Kelvin above absolute zero (−273 °C). If there is sufficient mass present within the cloud, gravity may overcome resistive forces and cause the cloud to collapse. Resistance against collapse is provided by the internal kinetic energy of the gas and dust particles and by the interstellar magnetic field that permeates the galaxy. Once resistance is overcome, gravity collapses the cloud until it has become hot and dense enough inside to begin fusing hydrogen to helium. At this point a star is born.

Initial Conditions

Each nebula has a minimum mass that determines whether collapse will occur or be retarded by other resistive forces. This mass is called the Jeans mass, named after British physicist James Jeans. Jeans reasoned that a cloud would collapse only if it lacked sufficient internal support against gravity. Clearly the criteria will vary from location to location in the galaxy.

In addition to the effects of magnetism, chemical composition plays a crucial role in the formation of stars. Nebulae composed of pure hydrogen and helium tend to remain massive as they collapse, failing to fragment effectively. The reason for this is a little complex, but note that a large volume of gas has quite a significant amount of internal energy – even if the temperature of the material is very low. As the gas collapses gravitational potential energy is converted to thermal (heat) energy, and this thermal energy causes the gas to resist further collapse. Cooling in primordial gas clouds is, therefore, a relatively inefficient process driven primarily by the formation of diatomic hydrogen gas (H_2). Consequently, the nebulae remain fairly massive even as they collapse, and they fail to fragment into smaller, denser and less massive cores. Thus the stars produced from this material tend to be particularly massive, and modeling suggests that this was an important factor in the formation of the first stars in the universe.

Subsequent populations of stars were polluted with elements heavier than helium. These metals are able to efficiently cool clouds of hydrogen and helium, and nebulae tend to fragment into smaller pieces. Metal atoms, and their associated ions, have a large number of potential energy levels associated with the larger number of electrons these elements possess.

If an electron orbiting the nucleus picks up a photon of energy at a short wavelength, the atom absorbs some of the ambient energy within the cloud. However, the electron can subsequently release this energy by dropping down to its original energy shell in smaller, less energetic hops. If the nebula is transparent to these longer wavelengths the energy escapes the cloud and the internal energy is reduced. This allows the cloud to shrink further. The combination of cooling, gravitational collapse and the conservation of angular momentum then dictates how the cloud shrinks and fragments.

With greater cooling and fragmentation, the stars formed from metal-rich gas tend to be smaller than those formed from metal-free or metal-poor gas. However, this is theoretical, and it has to be said, poorly constrained by observation. In 2011 a star, SDSS J102915+172927, was discovered in the galactic halo with a low mass and an extremely low metallicity – less than 1 millionth the abundance of metal in the Sun. This EMP (extremely metal poor star) followed the earlier identification of HE0107-5240 in the early part of the last decade.

The presence of such low mass stars threatens the assumption that early stellar generations were more massive – or at least that there is a simple relationship between metallicity and stellar mass function. There was an assumption that interstellar chemistry had to cross some threshold value of metallicity in order to allow for the formation of low mass stars. However, observations of galactic halo stars are increasingly challenging this assumption. Moreover, observations of stellar populations at higher red-shift, i.e., greater age, don't show significant differences in the stellar IMF with increasing age or decreasing metallicity. Finally, observations of metal-poor Population II stars show little evidence for the kinds of explosions predicted for the first metal-free stars. Instead most closely match the elemental abundances seen for conventional core-collapse supernovae, or those with substantial fallback into a central black hole, and with substantial mixing within the stellar envelope. However, the number of very low metallicity stars observed is marginal, and it remains possible that we are only seeing the effect of a few massive stars that are not representative of Population III as a whole.

If the effect of metallicity on stellar birth produces a mixed signal, it appears as though this factor does affect the later evolutionary stages of the star. In this chapter we shall examine how metallicity may influence some particularly lively stellar fireworks.

It is generally agreed that the vast majority of stars form in clusters and associations – looser aggregates of stars. Each cluster may contain up to tens of thousands of members with a few clusters attaining membership in the hundreds of thousands. The vast majority of stars that form are low mass (less than the mass of the Sun), with only a sparse number of massive stars in association with these. In general this picture holds across all of observed interstellar space. The reason for this pattern of star formation is unclear, but it appears to hold across nebulae of varying composition.

A further clear pattern emerges in the distribution of stars that form. The most massive stars form from the densest regions of the parental nebula, with lower mass stars forming further out. Furthermore, in general massive stars are rare to non-existent members of low mass clusters, only appearing when the mass of the cluster as a whole is relatively large. However, there was thought to be a problem with this simple process of massive star formation. Massive stars produce copious radiation and intense stellar winds, and the expectation emerged that these processes would limit star formation by accretion to stars with masses less than a few times that of the Sun. It was assumed that massive stars were formed via collisions between low or intermediate mass stars. This merger model certainly could reproduce massive stars formation, but is it their sole method of creation?

Observations by Henrik Beuther (University of Heidelberg) and Peter Schilke (University of Cologne) of a star-forming region IRAS 19410–2336 revealed that the process of massive stars formation appeared to mirror that of lower mass star formation with a very similar scaling of the mass of protostars. The observations were done in the millimeter (microwave) range where the dusty material comprising the nebula is transparent. This allowed detailed observations of the internal structure of the nebula.

Beuther's and Schilke's work revealed a nebula containing denser portions called cloud cores. Embedded within these cores lay still denser clumps of gas in which the massive protostars were forming, in a hierarchical arrangement akin to a Matryoshka doll. The nebula contained smaller cloud cores, and within these lower mass stars were presumably forming. The process appeared identical for massive protostars and their lesser cousins. Thus, although the merger model is not excluded, these observations suggested that mergers were not necessary to form massive stars.

Mark Krumholz (University of California, Santa Cruz) carried out another study, published in 2009, which detailed in three dimensions the formation of massive stars. This was no mean feat, particularly as earlier, simpler models failed to show accretion by protostars with masses exceeding 20 times that of the Sun. Again the culprit was radiation pressure. At this mass, the radiation produced by the protostar overwhelms the force of gravity, halting further accretion. However, these earlier models assume spherical accretion onto the surface of the protostars, when we are already aware that stars form at the heart of spinning accretion discs.

Modeling in three dimensions took 40 days of computing time on 256 processors. The results were astonishing. Mark Krumholz's model described the first 50,000 years in the life of a protostar. During the first 4,000 years the cloud collapses into a thick disc with the protostar at its heart. During the ensuing 20,000 years, spirals within the disc allow the continued accretion of material onto the protostar. At 17,000 years radiation pressure begins to affect accretion. However, radiation primarily escapes from the rotation poles of the protostar while accretion continues along equatorial regions. By 50,000 years three protostars complete formation with masses of 20, 37 and 44 times that of the Sun. The key to the success of these models was the consideration of accretion in three dimensions. Accretion primarily occurs along the equatorial plane of the protostar, driven by gravity, while winds blow material outward along the polar axis of the protostar.

Despite the success of these models and observations we shouldn't completely discount the role of collisions between protostars. Modeling of star formation revealed a frenetic process of protostar interaction, an occasional merger and, on occasion, an untimely ejection. Particularly dense clusters of stars, those formed from larger and more massive nebulae, seem prone to stellar mergers. The question is how much of this mass these stars can hold onto. At present models tend to show even the most massive star (up to 1,000 times the mass of the Sun) will still be pared down to more modest dimensions by stellar winds. However, much work remains to be done with the influence and extent of stellar mass loss; thus, this remains an open book awaiting further discovery.

The Life of a Star

Once a protostar has formed, gravity continues to contract the object, heating it steadily until nuclear reactions can begin. The lowest mass stars take over 2 billion years to stabilize, but as the mass increases, the time spent growing and maturing decreases. A star with the mass of the Sun spends about 30 million years descending from a nebula, through its protostar phase to the main sequence. But something as massive as 20 or more times the mass of the Sun will spend less than 50,000 years completing this journey. As the core increases in temperature, first deuterium, then lithium is consumed. These are mere burps along the road to stardom and have little or no influence on the process of massive star formation.

Hydrogen will initially fuse through the proton-proton (ppI) chain, but as temperatures continue to climb the CN cycle will engage, driving the production of most of the stellar energy. Once radiation and stellar winds have dispersed a critical mass of material around the star, accretion ceases and stellar winds rip free in all directions, clearing the remaining gases away.

Massive stars are born with high surface temperatures, and the radiation they release is mostly in the form of ultraviolet. This ionizes the remaining gases around the star, forming an emission nebula and revealing the birthing chamber in all its splendor.

Around the brilliant core of the stellar cluster lies those lesser stars, still condensing from the remains of the cloud. Many glorious examples of these nebulae are known, the nebula in Orion perhaps being the most famous. Around the central Trapezium cluster of O-class stars exists a multitude of protostars still surrounded by discs of gas and dust. Many show the impact of radiation from the central Trapezium cluster, with bow shocks around their leading edges and ionization of the gases within the disc.

One wonders if the Sun formed in such circumstances. The nascent Sun descended, contracting and heating, with Jupiter forming from within the swirling disc of gas and dust surrounding it. But before the Sun ignited its engines, a nearby star evolved away from the main sequence, its high mass dictating a short life. The Sun and infant planets breathed a temporary sigh of relief as the ultraviolet radiation that threatened their formation died away. A million years after this star left the main sequence, cooling and brightening as a red supergiant, its core collapsed, triggering a supernova. The shock wave battered the infant Sun but showered it with a glittering gift of radioactive elements that helped heat and differentiate the planets as they condensed around the Sun.

Billions of years later, those distant memories have long since faded, but left their footprint in the ratio of isotopes of magnesium in the rocks that form our terrestrial worlds. Core collapse supernovae make abundant aluminium-26, which decays in a few hundred thousand years to form magnesium-26. Many of the asteroids in the Solar System are rich in this isotope. And since aluminium-26 has a short half-life the only way it could have acquired it is by close proximity to a supernovae. By implication, many other useful elements such as the oxygen in our air, water and rocks; the iron in our blood; and the calcium in our bones may also have come from this same death in the neighborhood.

We may owe much to the death of a massive star.

The Main Sequence

A star is born when the collapsing and heating object reaches the main sequence. The starting position on the main sequence is called the "zero age main sequence," ZAMS for short. Here, the energy produced by nuclear reactions balances the inward pull of gravity. Contraction ceases, and nuclear reactions between hydrogen nuclei provide the necessary energy to stave off gravity's fatal attraction. Stability is achieved. The period the star spends on the main sequence depends critically on the initial mass of the star. This is detailed in the table below.

Stars entering the main sequence emerge at a location set by their mass and to a lesser extent by their chemical composition. Metal-poor stars are found on the sub-dwarf branch (luminosity class VI), but we can effectively ignore them for now because the metal-poor stars in our galaxy are all of low mass. The stars that form the bulk of the thin disc stars in our galaxy and those stars forming today in other parts of the universe fall onto the main sequence (luminosity class V). These stars form a sequence that follows the so-called mass luminosity relationship (Table 2.1).

In outline, a simple relationship exists that links the mass of the star to its luminosity. The relationship works for all stars that fuse hydrogen to helium. However, the precise link between mass and luminosity varies as we ascend the main sequence, and the manner in which energy is created and transported varies (see Chap. 1). For stars with masses exceeding 20 times that of the Sun the relationship is a very simple one. As mass goes up luminosity goes up in turn. Beneath this, at the lower regions of the main sequence, the relationship varies considerably, with different expressions added to tweak the mathematics in line with observations.

The Main Sequence Lifetime

A star will live on the main sequence for as long as it has hydrogen fuel to support it. During its time here, the core converts hydrogen to helium, and the core slowly fills with spent fuel. For the lowest mass stars the helium produced by fusion is continually mixed into the bulk of the star by global convection currents. However, in stars with masses above approximately 0.25 solar masses, convection does not extend throughout the star, and the fuel available to the core is restricted at birth.

In a Sun-like star, with a radiative core, helium settles towards the center of the core throughout the main sequence lifetime. In more massive stars convection mixes the helium

Table 2.1 Main sequence mass and lifetime. One WD refers to a white dwarf with a core dominated by the elements oxygen and neon, while CO refers to a white dwarf with an interior dominated by carbon and oxygen. He WD is a helium-dominated white dwarf

Initial mass	Luminosity L/L. (ZAMS)	T (ZAMS) 10^3 K	Spectral class (ZAMS)	Absolute magnitude (ZAMS)	Time on main Sequence	Total lifespan	Fate (Dependent on mass loss)
120	5,000,000	53	O2	−12	3.0	3.9	Black hole
60	700,000	48	O3	−8.7	3.5	4.0	Black hole
40	400,000	44	O5	−5.6	4.4	5.0	Black hole
20	60,000	35	O8	−5.0	8.2	9.0	Neutron star
12	10,000	28	B0.5	−4.0	16	18	Neutron star
7	3,000	21	B2	−1.5	43	48	CO/ONe WD
5	1,000	17	B4	−0.8	94	107	CO WD
3	100	12.2	B7	−0.2	350	440	CO WD
2	25	9.1	A2	1.4	1.16 billion	1.36 billion	CO WD
1.5	7.5	7.1	F3	3.0	2.7 billion	2.9 billion	CO WD
1	1	5.64	G5	5.2	11 billion	12.3 billion	CO WD
0.8	4/5	5.3	K0	6.4	25 billion	27 billion	CO WD
0.3	1/100	3.5	M5	11.6	350 billion	400 billion	He WD

with the remaining hydrogen in the stellar core. However, the end result is a fall in the number of particles present in the core. Four hydrogen atoms have been converted to one helium atom. It should be said, though, that the term "atom" is a tad misleading at the very high temperatures found in the stellar core. All atoms of hydrogen and helium are fully ionized and so consist of free electrons and nuclei.

As the number of particles decreases the pressure provided by their constant movement within the core decreases, and the core is forced to contract to compensate. Gases that contract heat up, and so throughout the main sequence the temperature in the core of a star increases. This accelerates the rate of nuclear reactions, and the luminosity of the star slowly picks up.

Over the last 4.5 billion years the Sun has grown hotter and slightly larger, and is now more luminous in response to this effect; the same is true for more massive stars. As the fuel is consumed the luminosity slowly increases.

The effect is quite subtle in comparison to later changes, but over the main sequence lifetime, low and intermediate mass stars (0.1–1.5 solar masses) slowly migrate up the main sequence over a short distance before moving away from it at greater speed once the fuel has gone. More massive stars move only slightly upwards, but predominantly execute a straight and very rapid run to the cooler portion of the HR diagram before exploding. The difference in behavior reflects the manner in which fuel is consumed and the limited time the star has to respond to changes in fuel availability before other less appealing effects overwhelm the star's sensibilities.

In general the time a star has to spend on the main sequence reflects its mass – although chemical composition is also a factor. Low metallicity stars lose energy more readily to the outside universe and hence have to generate it faster to remain stable against gravity. Thus a low metallicity star lasts for a shorter time than a star of the same mass but with more heavy elements (metals). The basic idea is that the more energy a star gives out, the faster its fuel will be used up and its life terminated.

Thus massive stars live much shorter lives than less massive ones.

Instability on the Main Sequence

We like to think of the main sequence as a happy time for a star. It is producing energy that allows it time to breathe and hold off the debt collector, gravity. However, this cozy textbook picture can be misleading. For many stars processes occurring within the stellar envelope mean that these stars pulsate, and for a few massive ones there is a source on instability within the core.

Many intermediate mass stars fall inside the so-called instability strip – a zone on the HR diagram stretching from the cooler edge of class A through class F on the main sequence and white dwarf cooling tract to class G in the giants, which contains stars that pulsate in a regular manner.

Main sequence stars in class F (~1.25–1.8 solar masses) have one significant behavioral difference from their smaller and larger siblings. These stars are intrinsically variable. Variability is a consequence of changes in the physical behavior of the outer portion of the envelope of the star. In main sequence stars lying in this mass range, the temperature reaches 50,000 K just beneath the photosphere. At this temperature helium becomes doubly ionized (helium atoms lose both of their electrons). Helium atoms can lose two electrons with the second electron coming off as temperatures approach and then exceed 50,000 K.

As any chemist will tell you, removing electrons from helium is a difficult business that requires a lot of energy. Therefore, as temperatures approach those critical for electron removal, the ability of the envelope to lose energy falls away sharply. Available energy becomes trapped in the layer of ionizing gases. Moreover, doubly ionized helium is more opaque to radiation than helium in its singly ionized state.

As radiation becomes trapped in the layer the material expands, causing the star as a whole to inflate. Expansion causes the layer to cool, which reverses the process – electrons recombine with the helium. Opacity drops, radiation can escape and the layer relaxes once more. The ensuing contraction then reverses these steps and the process continues.

The incessant cycling in the ionization state of this relatively thin layer drives expansion and contraction of the envelope, and this, in turn, drives the observed changes in luminosity. These variations in stellar girth and hence brightness can only occur where the layer in which helium ionization is occurring has a particular density. This mechanism drives luminosity changes in Cepheid variables – the stellar cornerstone of much of modern cosmology, used to determine the distances to distant galaxies and hence the expansion rate of the universe.

In 1962 Robert Christy proposed that a variation on this mechanism – pulsation driven by the ionization of hydrogen – might explain cyclical expansion and contraction observed in giant stars such as S U Draconis. In Chap. 3 we will examine the potential impact of these processes on the evolution of the universe's most massive stars.

In stars hotter than class F helium II ionization occurs at too shallow a level to drive the pulsation in the manner described. Helium ionization in A- and B-class stars occurs near the photosphere, and insufficient energy can be stored within the envelope in these stars to drive pulsations. Too great a depth and the star will not pulsate; nor will those in which the mass is too high and the level of overall ionization occurs is considerably greater. The instability strip thus demarcates a zone of stellar mass and temperature where pulsations occur, driven by the ionization of helium. This strip extends from the white dwarf tract (ZZ Ceti variables), through the main sequence and up into the giant branch. Here stars with masses between 5 and 9 times that of the Sun appear as Cepheid variables when evolutionary processes carry them through the instability strip.

An interesting idea is that pulsations, driven by different ionizations, may drive the final changes in the life of low and intermediate mass stars. If you look at planetary nebulae, the complex inner nebula is often surrounded first by an inner layer made up of closely spaced shells, and this in turn is cloaked in a broader, messier outflow. These changes reflect alterations in tempo as the star approaches its final scene in the movie. Perhaps the earlier stages are associated with mass loss driven by the ionization of hydrogen, but as the star loses mass more and more energy can escape by radiation. The star then heats up, and the helium ionization regions approach the stellar surface. This change drives the appearance of more closely spaced shells, which ultimately give way to the final planetary nebula as the remnant star collapses what's left of its envelope onto the nascent white dwarf within.

β-Cephei Variables

Further up the main sequence are another set of variable stars, the β-Cepheids. These have masses between 10 and 20 times that of the Sun and are thus hot, blue, B-class stars with surface temperatures exceeding 18,000 °K. In these stars helium is doubly ionized throughout the bulk of the star, and this cannot provide the sort of pressure-valve mechanism seen in F-class instability strip stars. Instead these hot stars pulsate because iron and similar relatively massive elements acts as the valve. In a manner analogous to helium at lower temperatures, when iron is sufficiently heated electrons are driven off the metal (which is already partially ionized at these high temperatures). Progressive ionization of this iron then allows energy to flow through from the interior to the exterior of the star, and the star can breathe a sigh of relief, contracting inwards. However, once the energy has passed through and the layer containing iron ions cools, some of the liberated electrons recombine with the ions and trap energy once more, causing renewed expansion and the pattern to repeat. Many O and B stars show this pattern of variability, and it is proving rather useful to astronomers (Fig. 2.1).

As stars pulsate the pattern of pulsation reflects their internal structure. Pulsations generate internal waves that course through the star like giant earthquakes. And just like on Earth, seismic waves can be used to investigate the planet's deep interior. Astronomers have used this technique, known as astroseismology, to probe the interiors of many stars, the Sun included. For example, astroseismological observations of the ten solar mass B-star HD129929 and the 20 solar mass O-class star HD46202 reveal that convection within the heart of these stars overshoots the edge of the core. This implies that additional hydrogen from the base of the stellar envelope could be brought down into the core, affecting the core's mass and potentially lengthening the time the star has to fuse hydrogen – the star's main sequence lifetime.

In still more massive stars, a second mechanism comes into play that generates pulsations within the star. The so-called epsilon (ε)- or Eddington mechanism was first evoked in the 1930s to explain the pulsations of Cepheid variables. However, subsequent work indicated that the kappa mechanism described above was the

The Anatomy of Stellar Life and Death   53

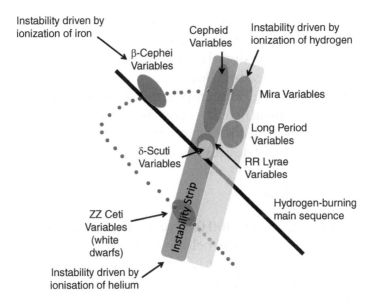

FIG. 2.1 Variable stars and the HR diagram. Stars in the instability strip pulsate because of ionization of helium just under the surface of the stars. β-Cepheids pulsate because of radiation periodically trapped and released by a layer of partly ionized iron some distance under the surface of the star

root cause of Cepheid instability. However, in stars with masses in excess of 60 times that of the Sun, a type of instability associated with the production of nuclear energy sets in. This epsilon mechanism causes small but noticeable changes in the luminosity of massive O and hydrogen-deficient Wolf-Rayet (WR) stars.

In 1992 it was reported that the WR star WR40, or HD96548, pulsated with a period of 627 s (or a little over 10 min). This period was in keeping with theoretical predictions of pulsations driven by the ε-mechanism. Of importance, although the change in magnitude was slight (0.005 magnitudes), this type of pulsation may be an important factor in driving mass loss in these very massive stars. In turn a simple pulsation may affect the type of death the most massive stars can undergo.

Therefore, although we like to think of the main sequence as the quiet time for a star, many stars, and probably all of the most massive stars, undergo pulsations that affect them and potentially foreshorten their lives. The main sequence is not so quiet.

Throughout the main sequence massive stars rapidly deplete their inventory of core hydrogen by the CN cycle. At somewhat higher temperatures hydrogen is also consumed by a final ring of reactions called the CNO cycle. This succession branches off the CN cycle, producing a slightly different set of isotopes, but ultimately helium is the end-product.

As the abundance of hydrogen falls to less than 1 % of the mass of the core, insufficient energy is generated by fusion, the temperature gradient declines and convection ceases. All remaining hydrogen is rapidly destroyed, and the core begins to contract.

As Above, So Below

Over the ensuing 50,000–100,000 years, depending on mass, the star expands and cools as the helium-rich core contracts and heats up. Luminosity remains roughly constant for these massive stars, during this transit, which is in stark contrast to Sun-like stars, which steadily grow brighter.

The dichotomy is explained by the behavior of the hydrogen-free core. Stars like the Sun consume only about 10 % of their mass during the main sequence phase. The helium core that is left over sets the luminosity of the star. As the Sun's core contracts, hydrogen fusion switches on in a shell around it, adding fresh helium to the core and thus increasing its mass. As the mass increases so does the luminosity of the star. These stars thus ascend the red giant branch. They do this quite slowly over hundreds of millions of years. All stars with masses of less than 2.25 times that of the Sun follow the same route and pile up at the same location on the HR diagram. This location is set by a mass of helium roughly half that of the Sun (0.5 solar masses). At that point helium fusion sets in, and the stars depart the red giant branch for a few tens of millions of years.

Massive stars (indeed all stars with masses exceeding 2.25 solar masses) already have helium core masses exceeding 0.5 solar masses. Therefore, as their helium cores contract and heat up, the concomitant expansion of their hydrogen envelope drives a cooling trend on the HR diagram but not an increase in the brightness of the star, its luminosity. These stars merely head to cooler climes before helium fusion sets in and the star embarks on the next phase of its life.

During this first expansion of the star, helium can be dredged out of the core as the zone in which convection is occurring in the envelope extends downwards. As it reaches what was the edge of the former core helium, along with isotopes of nitrogen, appear at the stellar surface. These elements were produced earlier by hydrogen fusion, through the CNO cycles. This, first dredge up as it is known, can be a useful diagnostic tool allowing astronomers to tell that the star has moved on from the main phase of its life to something a little more dramatic.

The transition to giant-hood is brief, which accounts for the scarcity of stars between classes O and B, and the cool red giant and supergiant classes. Stars spend so little of their lives here that they are unlikely to get caught in the act of transformation. The HR gap is thus a conspicuous feature of the HR diagram but one that is readily explained by models of stellar evolution.

Helium Ignition and Subsequent Evolution

Helium fusion kicks in with relative passivity in massive stars. Once the helium core has heated to temperatures in excess of 100 million degrees Kelvin, it ignites gently. At this point helium begins a frantic three-way fusion reaction that produces first carbon and then oxygen. The triple-alpha reaction, as it is known, releases roughly one-tenth the energy produced by hydrogen fusion. All subsequent phases of fusion release proportionately less energy in turn. Why?

Nuclear fusion converts a small fraction of the mass of the parent nuclei into energy, via the eponymous equation $E = mc^2$. The energy liberated is called binding energy, and it is a fraction of the energy in the nucleus of the atom that links the component quarks together. Hydrogen fusion converts approximately 0.7 % of the mass of the four hydrogen atoms into energy. However, considerably less binding energy is liberated when helium fuses in threes (Figs. 2.2 and 2.3).

Indeed, close inspection of the diagrams here reveals that only one-tenth the energy is available to the star. One might then reasonably expect that helium fusion lasts for a tenth the time the hydrogen fusion did. After all, if a tenth of the energy is available then the star must use it ten times faster to get the same bang for

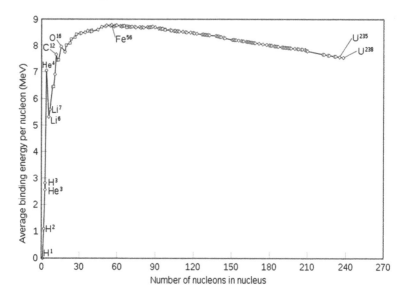

Fig. 2.2 Nuclear binding energy showing the change in binding energy with increasing mass of nucleons (protons and neutrons) in the nucleus of the atom. Hydrogen fusion to form helium releases approximately ten times the energy of helium fusion to form carbon (Image Source: Wikipedia Commons)

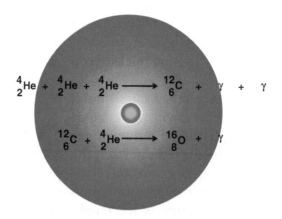

Fig. 2.3 The triple-alpha reaction. In evolved stars helium reaches temperatures in excess of 100 million Kelvin and fuses in threes to make carbon-12. At later times carbon-12 can pick up another helium nucleus and transform into oxygen-16. Each reaction releases gamma rays

its bucks. However, helium fusion is limited by one other factor: the speed of the reactions. Helium fuses to carbon and oxygen with even greater temperature sensitivity than hydrogen. A 2 % increase in temperature causes a doubling in the rate (rate varies with T^{40}). The very high rate of reactions means that nuclear fusion lasts only 150 million years or so for the Sun – compared with 11 billion years fusing hydrogen. For a massive star, helium fusion lasts between 100,000 and 1 million years, with a time that decreases with increasing mass.

For stars of between 0.5 solar masses (depending on the mass lost through stellar winds) to seven times that of the Sun, this is as far as nuclear fusion goes. The star has insufficient reserves of gravitational potential energy to raise its core temperatures high enough to fuse the carbon or oxygen produced by helium fusion. The star stops its ascent of the fusion ladder. Stellar winds then terminate the evolution of the star by the removal of the stellar envelope as a planetary nebula. The core shrinks and begins a protracted period of cooling as a white dwarf. This is the end for single, low and intermediate mass stars.

Stars with masses exceeding seven times that of the Sun generate core temperatures in excess of 800 million K. This is sufficient to allow carbon nuclei to combine in pairs directly to form neon. Carbon fusion produces even less energy than helium fusion and so lasts even less time. After only a few hundred years all the carbon is exhausted, and the star attempts to repeat the trick again (Fig. 2.4).

With the exception of a narrow window in mass of 9–9.25 solar masses, stars with masses above nine times that of the Sun initiate neon fusion and begin the final death-dive. Neon fusion lasts 12 months at most in stars with masses above 20 times that of the Sun and generates paltry amounts of energy, per unit mass, compared to the original stock of hydrogen. When that card has been played oxygen burning occurs for several months until silicon and sulfur are produced. Those stars in the range of 9–9.25 solar masses terminate fusion at the carbon-fusion stage, leaving either an oxygen-neon-magnesium white dwarf or they expire as supernovae. The latter is still a contentious issue, but is returned to in Chap. 6 (Figs. 2.5 and 2.6).

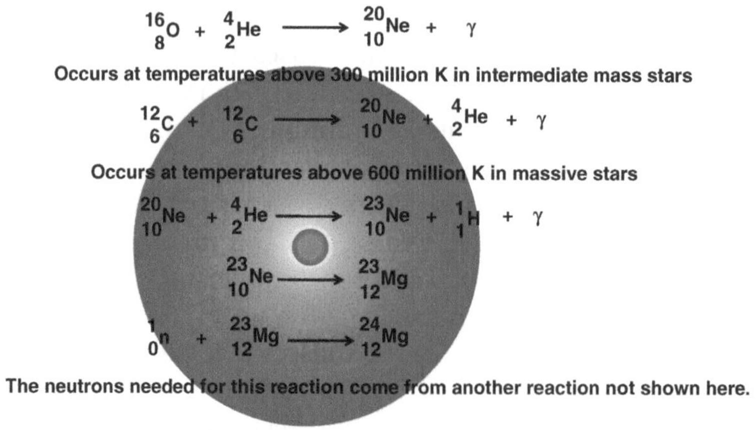

$$^{16}_{8}O + {}^{4}_{2}He \longrightarrow {}^{20}_{10}Ne + \gamma$$

Occurs at temperatures above 300 million K in intermediate mass stars

$$^{12}_{6}C + {}^{12}_{6}C \longrightarrow {}^{20}_{10}Ne + {}^{4}_{2}He + \gamma$$

Occurs at temperatures above 600 million K in massive stars

$$^{20}_{10}Ne + {}^{4}_{2}He \longrightarrow {}^{23}_{10}Ne + {}^{1}_{1}H + \gamma$$

$$^{23}_{10}Ne \longrightarrow {}^{23}_{12}Mg$$

$$^{1}_{0}n + {}^{23}_{12}Mg \longrightarrow {}^{24}_{12}Mg$$

The neutrons needed for this reaction come from another reaction not shown here.

Fig. 2.4 The fusion of carbon to make neon and magnesium. With the exception of the top reaction these occur in the cores of massive stars where temperatures exceed 800 million Kelvin. These reactions are fairly complex, with a variety of products that are ultimately dominated by neon with lesser amounts of magnesium

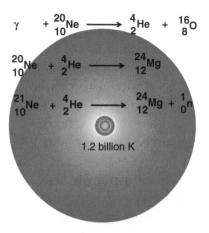

$$\gamma + {}^{20}_{10}Ne \longrightarrow {}^{4}_{2}He + {}^{16}_{8}O$$

$$^{20}_{10}Ne + {}^{4}_{2}He \longrightarrow {}^{24}_{12}Mg$$

$$^{21}_{10}Ne + {}^{4}_{2}He \longrightarrow {}^{24}_{12}Mg + {}^{1}_{0}n$$

1.2 billion K

Fig. 2.5 Neon fusion occurs at temperatures around 1,200 million Kelvin. The reactions are a little complex, but magnesium is the ultimate product, along with neutrons. These free neutrons are soaked up in the reaction shown in Fig. 2.4 that also produces magnesium in the cooler shell above this one

For the most massive stars silicon fusion represents the final desperate act. Very little useful energy is released, and within a day more than a Sun's mass of silicon has been turned unproductively into iron. Any attempt to fuse iron is futile. As temperatures rise,

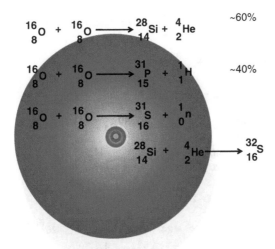

FIG. 2.6 Oxygen fusion. At temperatures around 1,500 million Kelvin oxygen fuses to make silicon and sulfur. The phosphorous shown here is converted into sulphur so that around 60 % of the core becomes silicon while the remainder is sulfur

iron doesn't fuse; it simply soaks up energy from the core. Were fusion possible, the outcome would be cooling of the core and the end of the reactions (Fig. 2.7).

Throughout the latter stages of nuclear burning the amount of energy released as ghost-like neutrinos accelerates. For a star born with 20 solar masses the luminosity in neutrinos rises by nearly one million fold from hydrogen to silicon burning. Most of these losses come from the creation and annihilation of positrons and electrons from energetic gamma rays. During the annihilation, some products end up as neutrino-antineutrino pairs rather than gamma rays. This is wasted energy that zips merrily out of the dying star. Consequently, the star is forced to contract its core further, raising the temperature and the rate of nuclear burning to compensate for the loss. But higher temperatures accelerate the process and a vicious, and self-destructive, cycle ensues.

Indeed, the mere act of pair production (electron-positron pairs) robs the star of some of its critical support, further hastening core collapse, temperature increase and enhancement in the rate of nuclear burning. For stars with less than 50 or so solar masses in their helium cores, pair production is not fatal in itself. However, later on pair production can have unfortunate effects for stars more massive than this limit.

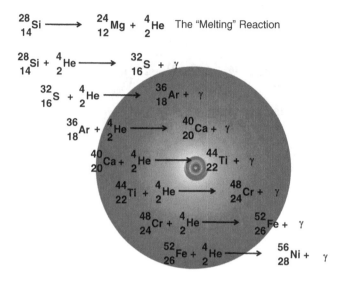

FIG. 2.7 Silicon fusion. Something of a misnomer, silicon fusion involves both its annihilation through the action of energetic gamma rays and a barrage of reactions where silicon serves as the scaffold for the assembly of iron and nickel. In the cores of massive stars energetic gamma rays break up silicon, releasing alpha particles. These are added to other silicon nuclei in a ladder of reactions that ultimately produce nickel-56. Silicon melts at temperatures above 2.7 billion Kelvin. Temperatures are higher in more massive stars and the resulting iron core can grow faster and larger before it implodes

Stages in the Life of a 25 Solar Mass Star

Burning phase	Required temperature (degree Kelvin)	Required mean density (gm per cubic cm)	Duration
Hydrogen burning	4×10^7	5	7,000,000 years
Helium burning	2×10^8	700	700,000 years
Carbon burning	6×10^8	200,000	600 years
Neon burning	1.2×10^9	4 million	1 year
Oxygen burning	1.5×10^9	10 million	6 months
Silicon burning	2.7×10^9	30 million	1 day

The increase in luminosity in the form of neutrinos provides the second explanation for the rapid decrease in the stellar burning time with each new fuel. The combination of lower productive energy release and the amount of total energy available causes the star to rush faster and faster headlong toward its grave. Each nuclear hit gives less return than the last.

As the last stages of nuclear burning are reached the interior of the star resembles an onion. The analogy with onions has become somewhat wearisome. Russian Matryoshka dolls are far more enlightening The inner iron core nestles within narrow shells of silicon and oxygen, which in turn snuggly settle within thicker layers of neon, carbon and helium, working outwards from the center. On the outside a layer of hydrogen persists – at least for single stars with masses of less than 25–30 times the mass of the Sun. Larger stars, or those in close binary partnerships, may shed their hydrogen (and sometimes helium) layers, exposing the hidden, former core.

Death of a Star

Once the core fills with more than 1.44 solar masses of iron the density and temperature reach a critical point at which energetic gamma rays begin to rip the iron nuclei apart. This process, called photodisintegration (meaning destruction by light), reverses the reactions that occurred over the preceding hundred thousand years or so. Iron becomes hydrogen, helium and neutrons. The next stage takes a lot longer to describe than it takes to happen. You may take a minute to read this paragraph, but in far less than a second the iron core is obliterated and its tattered remains implode.

During photodisintegration gamma rays split the nuclei down into smaller fragments; the process requires energy – the energy used by the star to support itself against gravity. In the final moments, as the soup of particles crushes inward at more than 70,000 km/s, electrons merge with the nuclei, further draining support from the moribund core. The products of this merger are neutrons and a sea of neutrinos that pour out of the core of the star, whipping away 99 % of the core's internal energy. However, the journey of these neutrinos, normally a peaceful and non-taxing ride, is severely hampered by the extreme densities of matter found

62   Extreme Explosions

Anatomy of a Proto-Neutron Star 0.2 Seconds After its Formation

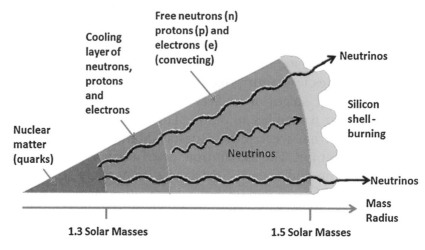

FIG. 2.8 0.2 s after the core collapses, the young, or proto, neutron star has a temperature in the hundreds of billions and is cooling quickly through the release of neutrinos. These particles can become trapped beneath the infalling silicon and oxygen shells and re-invigorate the stalled shockwave that lies near its outer surface 200 km above the center of the star. On the surface of the proto-neutron star silicon burning and other nuclear reactions continue

within the imploded core. Instead of wistfully departing the scene of the crime, they are briefly trapped within the carnage. It is this that leads to the next stage of the unfolding drama.

In the earliest models of core collapse, the formation of the neutron star was accompanied by a bounce as the material compressing into the proto-neutron star (as this hot, young object is called) overshot and bounced back outwards. However, it soon became clear that the energy of this bounce was insufficient to overcome the sheer weight of in-falling material in the overlying silicon and oxygen shells. The shockwave stopped in a fraction of a second, having traversed a pitiful 250 km or so from the ravaged surface of the proto-neutron star. Something else was needed to re-launch this failed shockwave (Fig. 2.8).

An initial solution to this conundrum came in the unlikely guise of the ephemeral neutrino. Neutrinos become shackled to the material within the dense stellar core. The neutron-rich matter forming the inner core boils at over 500 billion degrees Kelvin, and as the matter settles, neutrinos pour outward, impacting the collapsing silicon and oxygen shells surrounding the core.

Initially, in the 1980s, it was proposed that these escaping neutrinos would simply impart enough energy to the overlying silicon and oxygen shells and re-launch the stalled shockwave. However, the two-dimensional models used to justify this conclusion were too simplistic. Later three-dimensional versions of these seemed to falter.

In the early-1990s computer modeling by Adam Burrows (University of Arizona) indicated that the shockwave could be re-launched by the neutrinos but using a very different mechanism from a simple shove. Published in 1993, this model demonstrated that violent heating could drive convection-like plumes that ascended from above the edge of the neutron star – and they also dredged iron upwards from the deep interior to higher, cooler layers. Initially seen as a puzzling observation in SN 1987A, the presence of iron at early times indicated far more mixing between layers than was demonstrable in early modeling of supernovae. Thus convection seemed a reasonable starting point for an explosion, something that was also subsequently observed in laser-induced explosions in the lab.

However, these models were still done in two-dimensions, and it wasn't clear if they would apply to a more realistic three-dimensional structure such as a star. Further work with a class of explosions driven by so-called electron capture suggests that this mechanism might work for lower mass progenitor stars (9–9.25 solar masses). Electron capture probably drives the collapse of lower mass oxygen, neon and magnesium cores, rather than the more massive iron cores discussed here (Chap. 6). However, more recent work has indicated that the convection-driven mechanism is unable to drive explosions in more massive stars. There is simply too much mass in the overlying oxygen shell, and the process of convection delivers too little energy to drive it outward.

A further eclectic supernova model involved sound as the detonating mechanism. Again this work came from the University of Arizona, and the basic idea is fairly simple. Collapse of material onto the neutron star is an uneven and violent process; after all, material is imploding at 70,000 km/s. In this sonic model all the internal hullabaloo kicks off sound waves that rattle around in the collapsing core of the star, eventually causing the proto-neutron star to wiggle rather aggressively. Somewhat akin to watching the crowd at a 1970s punk concert, wavelike motion eventually sets in inside the material, which then initiates a shockwave that pushes out preferentially in one direction. Given the right amount of wiggling, the model indicated that the shockwave could drive violent expansion in one direction, and rather usefully also launch the neutron star through the expanding debris with a birth-kick – something that is observed in a number of supernova remnants.

However, is this real deal? At the moment no one is really sure. As yet we haven't exhausted all the possible players in this drama. As we already know stars rotate and possess magnetic fields. Indeed many of the sorts of B- and O-class stars that give rise to supernovae seem to possess relatively strong magnetic fields, and even those that do not should create such fields as the core of the star collapses inward.

Recall the physical property of rotating bodies: angular momentum. As a star's core collapses inwards angular momentum is conserved, and contraction must be accompanied by an increase in the rotation speed. Indeed, very basic calculations suggest that a freshly minted neutron star should spin at over 1,000 revolutions per second. Fast enough to dry your clothes in no time at all. However, neutron stars observed in supernova remnants spin nowhere near as fast as this. Typical speeds top out at a few tens of revolutions per second. For example the Crab pulsar spins at 33 times per second, and this is relatively nifty for a young neutron star. The question then arises, if they were born spinning much faster than this, where did all the extra spin go?

Astronomers have already observed young protostars ejecting material in jets and outflows from their rotation poles as they accrete material from their surroundings. This process allows the star to shed angular momentum effectively and collimate (or direct) magnetized outflows from the disc of matter surrounding them. In principle the proto-neutron star could do the same thing. Within the core of the star, matter accreting onto the proto-neutron star could spin it up and drive jets outwards along the star's rotation or magnetic axis. In addition, hot winds driven by nuclear reactions and neutrino emission within the accretion disc surrounding the proto-neutron star could also help drive material away from the collapsing core. Finally, the presence of a strong magnetic field within the proto-neutron star and any surrounding disc could launch jets of material outwards into the surroundings.

In reality all, some or none of these mechanisms could be at work, and it will undoubtedly take a lot of work and observation to discriminate between these models or their successors. Despite 50 years of work, in many ways the process of untangling the complexity of an imploding star is still in its infancy. However, one is reminded of Henry Louis Mencken's adage, "For every complex problem there is an answer that is clear, simple, and wrong." Keep watching for more complexity.

Fallback

In the outlined scenario, by some as yet unknown mechanism, a successful outgoing shockwave is launched. Within a day this penetrates the surface of the doomed star, scattering it to the four winds. However, what happens if the proto-neutron star doesn't impart enough kick to the material? In principle, if the stellar envelope is sufficiently massive, and particularly if it is densely packed, the outgoing shockwave may fail to puncture the star. In this case material would move outwards, albeit briefly; then either some or all of it would reverse its trajectory and begin raining back onto the newly formed neutron star. What would happen then? Work on so-called fallback has been modeled extensively over the decades, with some of the more recent work carried out by Stan Woosley (University of California, Santa Cruz), Chris Fryer (University of Arizona) and Andrew MacFadyen (New York University).

In stars with initial masses of over 30–40 times that of the Sun, stellar evolution produces a hydrogen-depleted object called a Wolf-Rayet star. We have encountered these already, but a few basic properties are worth reconsidering. Unlike a classic red supergiant, these objects may have only 10–20 times the mass of the Sun, but because they have shed their hydrogen-rich envelopes, these helium or carbon and oxygen-rich stars are small, dense and very hot. Modeling suggests that when core collapse occurs in these dense objects the shockwave is insufficiently powerful to push aside all of the material within the star. Instead a sizable proportion (up to a few solar masses) falls back onto the neutron star, triggering its implosion. The outcome is a stellar mass black hole: an object with 5–15 times the mass of the Sun squeezed into an area several kilometers across.

What happens next is in the realms of theory and computer simulation, but the suggestion is that the black hole may then accrete the bulk of the star through an accretion disc while simultaneously launching jets outwards through what remains. These jets may be visible as X-ray or gamma ray bursts, visible over cosmological distances (Chap. 4).

Depending on the nature of the collapse, and the amount of energy imparted to the surrounding material before the black hole formed, in principle some of the star may escape the clutches of the nascent black hole and become visible either as a supernova or in the form of jets directed towards us. Theory suggests that the supernova may be faint if little material escapes the clutches of the black hole. However, at least in some instances supernovae generated this way may be bright enough to observe over large distances.

Perhaps the most famous of these was the relatively close by SN 1998bw, associated with gamma ray burst GRB 980425. This Type Ic (hydrogen and helium deficient) supernova was particularly energetic, producing a disproportionately large quantity of radioactive nickel and a strong radio signal. The large mass of nickel (0.5 solar masses) suggested that in this explosion rather a large amount of material near the core managed to escape the nascent black hole. Other black holes, such as that found in Cygnus X-1, appear to have formed without much in the way of an explosion. So it appears that the formation of a black hole may be a very capricious process leading to very different outcomes for the parent star.

Observations of extreme metal-poor stars (EMPs) within the galactic halo indicate that their chemical compositions match the expected debris from supernovae that have substantial fallback. Work by Hideyuki Umeda and Ken'ichi Nomoto suggests that the composition of some EMPs can be produced by Population III stars that die in supernovae with substantial mixing and fallback onto the central stellar mass black hole. These supernovae lose most of the iron group elements, silicon and much of their oxygen into the maw of the growing black hole, while carbon and the products of helium and hydrogen fusion largely escape into interstellar space. Although clearly the output of a computer simulation, it is interesting that the model reproduces observations.

The Neutron Star

The product of core collapse for stars with initial masses of less than 25–40 times that of the Sun is an unusual object – the neutron star. Weighing in at roughly one and a half times the mass of the Sun but with a radius of only 10–15 km, it initially boils at over 500 billion degrees Kelvin. Its surface is a poor imitation of the original iron core of the once brilliant star, consisting of distorted iron group nuclei in a sea of electrons. Beneath this solid upper crust (perhaps 0.3–0.5 km thick) lies a 1–2 km lower crust of neutron rich material, primarily neutrons and electrons with some residual iron group nuclei. This layer lies atop a liquid mantle made of neutrons with a smattering of electrons and protons. Deeper still, within the central 3 km of the object, is a mysterious core consisting of neutrons, free quarks and gluons or maybe more exotic matter. The overall density is equivalent to, or exceeds that of, the nucleus of the atom – a stunning 100 trillion grams per cubic centimeter. One teaspoon would weigh in at roughly 100 million metric tons!

The gravitational field strength of this diminutive star is nearly 100 billion times that of Earth, and a spaceship taking off from one of these – were this possible – would need to reach one third the speed of light if it wished to escape the star's intense gravitational maw. Consequently, any object falling onto one of these objects releases rather a lot of gravitational potential energy upon impact. Gas, unfortunate enough to end up within the sphere of influence of a neutron star, impacts the surface, releasing a blast of X-rays.

Some unexpected observations were made of the central neutron star in the Cassiopeia A supernova remnant, observations best explained by nuclear reactions. The Cassiopeia A supernova remnant is approximately 300 years old, and the central neutron star appears to have a carbon atmosphere. Although it is possible that the carbon was accreted from the supernova remnant, it is unlikely that this would be the only element to be drawn to the neutron star. After all, hydrogen and helium are more abundant. Instead the suggestion was that hydrogen and helium, accreted from the surrounding remnant, underwent nuclear reactions and produced the carbon seen today. Observations of young neutrons stars would show if this was commonplace, as the reactions will release copious amounts of hard X-rays and gamma rays at characteristic wavelengths.

Aside from the high density and associated intense gravitational field, young neutron stars are intensely magnetic. A typical magnetic field strength is on the order of a trillion times that seen on Earth, or 10^{12} G. This field, coupled to the rapid rotation of the neutron star, readily accelerates particles embedded within the stellar atmosphere. Electrons spiraling within the magnetic field release primarily synchrotron radiation, but not exclusively at radio wavelengths. The 1,000 year old Crab pulsar also releases pulses of visible synchrotron radiation, and some, including the Crab pulsar, release pulses of X-rays. These pulses can be used to measure the rotation rate of the host neutron star, and rates have been found to vary from once every few seconds up to hundreds of revolutions per second for some old, "recycled" pulsars. Young pulsars, such as the Crab or Vela, spin at 10–30 revolutions per second. The Crab pulsar spins once on its axis every 33 milliseconds, and sensitive measurements confirm that it is slowly decelerating.

Why does it spin so quickly and why is it decelerating? Remember the conservation of angular momentum. As the core of the parent star shrank, its speed of rotation had to accelerate in order to conserve its momentum. Initially, this led to very high rotation rates, perhaps 200 times per second. However, as we saw, the braking of the magnetic field against the expanding debris in the supernova and the interstellar field leads to deceleration. A young neutron star decelerates by between 10^{-10} and 10^{-21} s for each rotation – not much, you may think, but after a million years or so this amounts to a few percent of the rotation of the star. With the declining rate of rotation falls the strength of the magnetic field. At some point the strength of the magnetic field falls below a critical threshold, and the star stops pulsing. It isn't clear entirely where this is, but it may lie somewhere near 100 million gauss, or approximately 100 million times the field strength around Earth.

Although neutron stars are born exceedingly hot, they cool quickly. Their surface area to volume ratio is high, but the density of the material resists cooling by standard electromagnetic radiation. Instead it falls to the elf-like neutrino to do the job of cooling the star. Over a few hundred years following inception, the temperature of the star falls from nearly one trillion degrees down to a few hundred million – a factor of 10,000. Meanwhile the surface temperature falls to around a few million by this point. The Cassiopeia A neutron star has already cooled to just over two million Kelvin within the last 300 years.

A recent paper by Dany Page suggested that the enhanced rate of cooling in this neutron star was due to the commencement of super-fluid behavior within the neutron star mantle. Super-fluidity is a property observed in some highly chilled gases on Earth. Despite the extreme heat and pressure, within the neutron star the neutrons can behave as a super fluid. Unusual quantum interactions between the neutrons allow them to move as though there is no friction between them. This allows convection to transport heat with high efficiency, thus cooling the core of the neutron star quickly. Although super-fluidity is predicted for neutron star interiors, exactly how this affects their properties is still in the realms of theory and speculation. The enhanced cooling rate seen in the Cassiopeia A remnant neutron star may be one of the first testable demonstrations of the role of super-fluidity within a neutron star.

A few neutron stars are born with peculiar magnetic fields that dwarf the heady strength achieved in more conventional pulsars by more than a factor of a hundred. These magnetars have fields so intense that iron nuclei in their crusts will be distorted into cigar shapes. Fields 100–1,000 times the strengths seen in a pulsar require an unusual mechanism to form them. The likely scenario involves a proto-neutron star spinning at more than 200 times per second. In this hot, fast rotating body a dynamo effect can develop that produces a field with a strength of 10^{15} times that of Earth. Such magnetars have fields so intense that they rapidly brake the rotation to less than once per second. Although sluggish in terms of rotation, the magnetic energy contained within these strange stars can warp and crack the surface of the neutron star, leading to the release of energy in a massive starquake.

Visible as a blast of gamma rays across galactic distances, these magnetars appear as soft-gamma ray repeaters, SGRs for short. Only a handful are known, along with a similarly paltry number of related but lower energy anomalous X-ray repeaters. Quite why some neutron stars are born as magnetars is unknown, but this may come down to the rotation of the parent star or perhaps even the mass of the star that gave rise to them. Certainly, to form a magnetar it seems likely that they must have inherited excess angular momentum from their parent star – but quite how this will relate to the properties of the progenitor are still unclear.

Magnetars may hold the key to some of the universe's most brilliant supernovae. We shall return to these eclectic beasts in Chap. 9.

The Fate of the Surrounding Star

What of the parent star itself? We already know it's doomed, but how does its demise demonstrate itself to the outside universe?

Within the immediate vicinity of the core, neutrons and other fragments of atoms punch into the surrounding silicon shell. Violent nuclear reactions generate a large quantity of radioactive iron, cobalt and nickel. In addition to these atoms, somewhere in this mêlée the soup of particles generates every other element in the Periodic Table upwards, through lead and uranium, towards an unknown summit. Above this, as this shockwave moves outward, it compresses and heats the various shells of material triggering a final wave of nuclear reactions.

For Type Ic supernovae, the shock wave finally leaves the star at the surface of the carbon shell, or a short way through an outer carbon-enriched helium layer after a few hours. These stars are relatively small, perhaps the diameter of the Sun. For Type Ib supernovae, the star is larger, perhaps a hundred times that of the Sun. Here, the shockwave continues expanding outwards until it exits through the more massive helium shell. And finally, for the bulkiest (if not the most massive) red supergiant stars, the shockwave ploughs onwards at 7,000–10,000 km per second before, perhaps a day later, leaving the hydrogen-rich outer envelope of the star. The progenitors of these Type II supernovae are perhaps 500–1,000

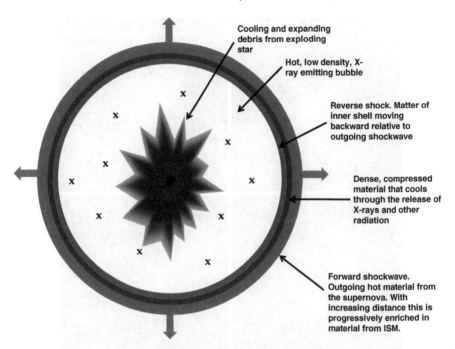

FIG. 2.9 The structure of a supernova shockwave

times wider than the Sun, hence the lengthy delay between the inception of the shockwave and its exit from the star. Even at a nifty click of 10,000 km per second the shockwave takes many hours to blitz the star's outer layers. Conversely, the progenitors of Type Ic supernovae may be only a few times the diameter of the Sun and take correspondingly less time to reach the stellar surface (Fig. 2.9).

What would the emerging shockwave look like? Probably not a smoothly penetrating sheet, for a start. The shockwave will undoubtedly be very uneven, and if it is driven by jets from deep within the star, will have a very aspherical shape. There is ample evidence that a large proportion of core-collapse supernovae are aspherical, many bearing witness to the effects of jets.

As the shockwave passes through material it imparts considerable energy which heats and expands the material. As the shockwave approaches the stellar photosphere, heating raises the temperature to hundreds of thousands of degrees Kelvin causing the material to emit copious amounts of X-rays and ultraviolet light (Fig. 2.10).

72 Extreme Explosions

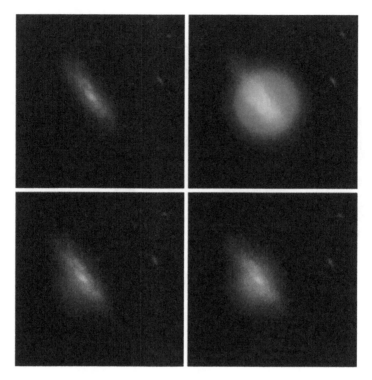

FIG. 2.10 Shock breakout in SNLS-04D2dc, detected as an ultraviolet flash by GALEX. (Images courtesy of NASA/HST/COSMOS/GALE)

In 2008 the Supernova Legacy Survey detected an explosion dubbed SNLS-04D2dc (Fig. 2.10). With some careful detective work the team combed archived ultraviolet images gathered by GALEX (NASA's Galaxy Evolution Explorer) taken during the previous few weeks. Diligence paid off, and a series of images were obtained that rev-ealed the death throes of a red supergiant as the supernova shockwave punched through the surface of the star and out into the giant's extended atmosphere. The supernova was an unremarkable Type II-P event, but this was the first time the shockwave had been seen punching out through this type of star.

A few months previously a similar event in a more compact progenitor was observed. This was associated with the Type Ibc supernova, SN 2008D. A Type Ibc explosion is one that initially displays helium in its spectrum, but later morphs into a helium-free event. Shock breakout through this compact star generated a pulse of X-rays detected by NASA's SWIFT telescope. The detection of this breakout, associated with this supernova, was serendipitous – a chance discovery, one which may not easily be repeated. The Supernova Legacy work was more akin to detective work but with a bit of luck in that the GALEX happened to be looking in the right place at the tight time.

Formation of Supernova Remnants

Following shock breakout the expanding debris immediately begins to cool. Energy is provided by the decay of elements synthesized in the explosion and, in some instances, by the interaction of the expanding debris with any surrounding material. However, it is worth considering the explosion itself, for it is very easy to misunderstand what has just happened to the material in the star.

It is easy to imagine that the shockwave, moving at over 10,000 km per second, simply lifts material ahead of it. However, instead it is a little bit more like the wave moving through a stadium crowd. The shockwave moves through and accelerates the bulk of the material but to a speed less than the wave as a whole is moving. The material accelerates to above the escape velocity of the star, but the shock wave is constantly moving outwards ahead of this mass of accelerated material. As a result, the shockwave moves away from the star, interacting with anything in its path, compressing and accelerating it. Lagging increasingly far behind is the expanding debris of the star. A small fraction of the star (less than 1 solar mass) moves bodily with the shockwave, out from the stellar surface and off into interstellar space, with the bulk of the material following more slowly behind this wave.

This can be seen in recent Hubble images of the SN 1987A supernova remnant (below). Behind the outgoing shock, expanding material, encountering the shockwave, generates a reverse shock that propagates backward, relative to the outgoing forward shock. This reverse shock slows the expansion of the remnant and can contribute to fallback onto the central object left by the explosion (Fig. 2.11).

After the shock has broken out in a typical Type II supernova, the expanding mass of debris brightens as its surface area expands. This phase lasts for perhaps 1 month. After this, cooling

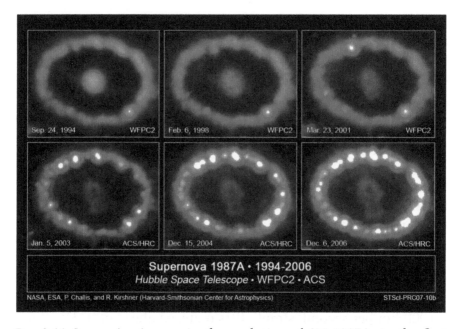

FIG. 2.11 Successive images in the evolution of SN 1987A. In the first image (*top left*) the shockwave from the explosion lies between the central *purple* supernova and the ring of material ejected prior to the supernova. By the time the *lower left* image is taken the shockwave has begun a strong interaction with the ring of material, but the supernova remnant (*purplish in color*) is clearly lagging behind (Image source: Hubble/NASA.)

begins to offset expansion and the luminosity falls. When the visible surface of the supernova has expanded to approximately 20 billion kilometers it becomes transparent, and the light produced drops dramatically. Plotting the visible light using photometric techniques captured from the explosion allows astronomers to construct light curves that provide valuable clues to the inner workings of the explosion. A light curve is usually a plot of visible magnitudes below maximum light versus time measured in days (Fig. 2.12).

During the first 25 days or so, after peak brightness, luminosity typically falls by 0.008 magnitudes per day on average. However from day 25–75 many Type II supernovae display a plateau in their light curve and are known as Type II-P explosions as a result. During this curious phase, hydrogen, abundant in Type II events, is initially very hot (more than 7,000 °K). However, as it cools

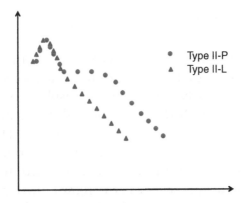

FIG. 2.12 A comparison of the light curves of Type II-P (Plateau) supernovae and Type II-L (Linear) supernovae

electrons can recombine with free nuclei and release a photon of light. This process continues until the temperature has fallen to less than 4,000 °K and the hydrogen is fully recombined as neutral gas. As the material is heated by radioactive decay internally, this process begins from the outside and works its way inward until the hydrogen is exhausted. However, this inward shrinkage is balanced initially by the expansion of the stellar debris, thus producing near constant light during the process. This explains the peculiar curve.

Stars with proportionately smaller (less voluminous) hydrogen envelopes will have smaller plateaus until this part of the light curve disappears altogether in stars with little or no hydrogen. Supernovae that display hydrogen but only in small amounts may appear as so-called Type II-L, or linear supernovae, as their curves are not punctuated by the plateau. Type II-L explosions consequently show more rapid decays in their light curves (around 0.012 magnitudes per day). In some instances limited amounts of hydrogen cause the supernova to transform from Type II to Type Ib events, and SN 1993J is a well-characterized example of this. Analysis later confirmed that SN 1993J has a massive, close companion star that appears to have removed much of the supernova progenitor's hydrogen envelope immediately before the explosion.

Messages From a Retreating Front

An informative feature of supernovae is the manner in which the expanding remnant undergoes various transitions as it expands and cools. At very early stages the photosphere of the supernova is moving outward with the shock-heated debris and illuminates only the outermost portion of the shattered star. However, as the material cools the photosphere, the virtual surface from which light escapes retreats inside this cocoon. This reveals progressively deeper and deeper layers as the light-emitting surface of the photosphere retreats further and further inside the expanding remnant. Elements forged at every layer of our broken Matryoshka doll appear, one after another as the light from the retreating front is absorbed by material lying ahead of it. Thus the interior of the star is opened much like peeling layer after layer of an onion. Not only does this reveal the inner architecture of the star but it can reveal how the star blew up. Close scrutiny of this process has allowed astronomers to understand how much mixing occurs inside the doomed star as the shockwave punches through the star. Such examination of SN 1987A showed that nickel-rich debris from the stellar core was present at far higher levels than expected, indicating thorough mixing.

Following the plateau phase the supernova continues to fade, but the decline is initially dictated by the decay of radioactive nickel-56 and cobalt-56, and much later by the decay of longer-lived radioactive isotopes such as titanium-44. Nickel-56 is abundantly synthesized in supernovae, as we have seen, but decays very quickly, with a half-life of 6.1 days to produce cobalt-56. This latter nuclide decays with a half-life of 77 days, and the decline in output from it principally governs the terminal decline in supernova luminosity. Decay of these nuclei is by positron and gamma emission. The primary gamma rays, and those produced by the annihilation of positrons, heat the expanding stellar material, causing it to glow.

Typical yields of radioactive nickel are on the order of 0.07 solar masses, but some core-collapse supernovae produce rather a lot more than this. SN 1998bw produced an estimated 0.5 solar masses; a typical Type Ia supernova produces 0.7 solar masses of radioactive nickel. These numbers seem rather abstract, so it's better to put them another way. If we go for the figure of 0.07 solar masses, typical of a Type II-P event, this is equivalent to 1.4×10^{29} kg – or 50,000 times the mass of Earth, give or take a few Mount Everests. Remember this is all in the form of a single radioactive element that is decaying and releasing energy.

SN 1987A, despite more than a few well-discussed oddities, beautifully confirmed many of the theories of supernovae. SN 1987A was peculiar because the progenitor was a compact blue supergiant star rather than a red supergiant, expanding if not contradicting the initial models of the day. The compact nature of the progenitor produced a slightly distorted light curve that essentially rose to a plateau and then declined, rather than peaking first. More energy went into detonating the star, rather than heating and driving material outward. Perhaps most significantly, a pulse of neutrinos was detected at the time of the explosion, confirming that a neutron star had formed. Finally, the production of radioactive elements by the explosion was observed at very early times, confirming models of nucleosynthesis. As previously mentioned, SN 1987A also revealed that heavier nuclei from deep within the star could be mixed upwards in the explosion, appearing at earlier than expected times in the spectrum of the explosion. This may be evidence of jets or plumes directed outward from the core of the star during the earliest stages of the explosion. Clearly, more work remains to be done in this area.

A typical Type II explosion releases approximately 10^{46} J of energy. Almost 99 % of this is in the form of neutrinos, but this still leaves a whopping 10^{44} J in kinetic energy and 10^{42} J in the form of electromagnetic radiation. Although the electromagnetic energy is disperses fairly quickly, the kinetic energy is a lot harder to dispose of. Initially, the kinetic energy in the debris expands freely – although, as we'll see, Type IIn explosions are exceptions to this rule. A typical core-collapse supernova may shed 5–10 solar masses of material. However, the bulk of the kinetic energy is carried by a much smaller fraction of this (maybe 0.25 solar masses). This small fraction constitutes the outgoing shockwave that will readily sweep up any material it encounters.

Given the low densities of material in space, this process may take hundreds of years to plough up a significant mass of material. However, what is swept up is violently heated to millions of degrees Kelvin. Material in the shell is thus both highly ionized and is a copious source of X-rays. At the shock interface electrons in the gas are accelerated and release synchrotron radiation primarily in the form of radio waves. Type II supernovae (particularly Type II-P) are usually strong emitters of radio waves. Thermonuclear Type Ia supernovae tend to be radio emission-free.

Some of the super-heated gas within the shell is able to fall back and cross behind the shock, forming a low density hot bubble of X-ray emitting gas. Over time this bubble grows as the shockwave moves outward at thousands of kilometers per second.

Over the ensuing millennia, the shockwave expands and the amount of kinetic energy present per unit mass in the debris falls. Moreover, the amount of matter in the shell is increasing, which further reduces the share of kinetic energy between particles. The shell thus begins a protracted period of deceleration. As the kinetic energy falls, the temperature begins to fall, and the amount of ionization decreases. The shell begins to falter in its advance. At a critical point, tens of thousands of years after the initial explosion, the matter in the shockwave has insufficient energy to continue bulldozing its surroundings. At this point, the light begins to go out, and the shocked material merges with the surrounding interstellar medium (ISM). All that remains is a ghostly pall of metal-enriched debris, ready to collapse again to form a new generation of stars.

By contrast material within the shocked bubble has such a low density that it is very difficult to cool by recombination. Thus supernovae-driven bubbles can remain in place for millions of years and extend beyond 1,000 light years in diameter. In galaxies like our own, not only do supernovae enrich their surroundings chemically, they mix and expel copious amounts of material from the galactic disc into the halo. In a large galaxy like ours, this is of only short-term interest, as this material remains trapped by our galaxy's gravitational field. However, in small galaxies, supernovae can drastically alter the fate of the galaxy by driving star-forming material out of the galaxy into intergalactic space. This can lead to the cessation of further rounds of star formation and end what was a fertile period in the galaxy's history.

In gas-poor elliptical galaxies, supernovae may also drive a galactic wind, whereby the shockwaves from explosions expand freely in the gas-poor medium and push outward into intergalactic space. Such activity may explain, in part, why elliptical galaxies support such feeble rates of star formation.

Conclusions

We've explored the basic ingredients for massive star formation and death and broached upon the areas where supernovae may affect galactic environments. With a basic foundation in supernova biology we are now ready to embark on the main focus of this book: the life and death of the most massive stars in the universe and some of the more enigmatic explosions stars produce when they die.

Part II
A Walk Across the Rooftops

3. Stellar Evolution at the Summit of the Main Sequence

Introduction

The fate of the most massive stars is an area fraught with uncertainty and is not without controversy. The stakes are high when it comes to this area of astrophysics, as some of the universe's most exciting and violent phenomena are associated with these rarest of objects. Like ephemeral diamonds, these singular stars provoke intense emotions among those who research them, emotions that can flare like the stars themselves. Consequently, this is a tricky chapter to write if one is to cover the current research adequately and without favor or bias (Fig. 3.1).

The main reason for the controversy is the manner in which these stars live, for it is within the manner of their lives that their ultimate fates are described. All stars lose mass as they age. For a star like the Sun, it will shed one hundred trillionth of its mass every year. Even over 10 billion years, this amounts to a careless whisper, one that is soon lost upon the eternal breath of the cosmos. However, as we ascend the main sequence radiation pressure from starlight plays an increasing role in driving up the rate of mass lost in stellar winds.

Why is this important? If a star loses a lot of its mass before its core implodes the outcome of this local catastrophe will change considerably. At death, a very high core mass might lead to the formation of a black hole or an explosion called a pair-instability event. Increases in the rate of mass loss would, however, lead to the more mundane formation of a neutron star, limiting even the most massive stars to a fairly predictable death.

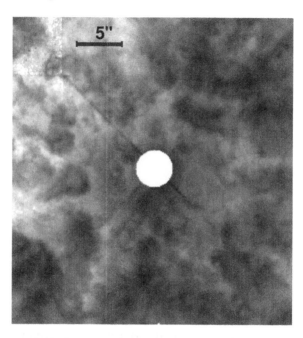

Fig. 3.1 Clumpy wind surrounding WR 124 HST image of surrounding nebula M1-67 with WFPC2. Possibly the first image of clumpy wind ejected by a hot central star (From ApJ vol 506, L127-131 1998 Yves Grosdidier and Anthony FJ Moffat)

Looking Deeper into the Controversy

Two quotations come to mind when considering this problem. In the *Astrophysical Journal* in 2003, Alex Heger wrote: "No two groups presently agree, in detail, on the final evolution of any massive star (including its explosion energy, remnant mass, and rotation rate) and the scaling of mass loss with metallicity during different evolutionary stages is widely debated." After nearly 10 years, although observations have begun to constrain some of these uncertainties, there is still widespread debate about the nature of mass loss in massive stars and the impact this has on their subsequent evolution and fate. Astrophysicist Nathan Smith commented in *Sky & Telescope* in 2011, "Stellar evolution models are based on a lot of assumptions about how stars lose mass during their lives. We have learned recently that some of the key assumptions are wrong at a basic level, and that stellar evolution models need to be rethought."

Although written nearly 10 years apart, it is notable that the only area of real consensus is that there is none – an oxymoron if ever there was one. There is still little overall consensus on the processes that determine mass loss in the universe's most massive stars. Consequently, tying down the fates of the most massive stars is proving irksome to say the least. The life of the most massive stars is limited to a few million years – five million is the most likely upper limit for stars with masses exceeding 40 times that of the Sun. However, calculated mass loss rates, even those based on the observations of the handful of ultra-massive stars that are known, carry large ranges and uncertainties. A star born with 150 times the mass of the Sun could end up pared down to a 25 solar mass Wolf-Rayet in its lifetime, or it could hold on to 100 solar masses, or more. The figures are not consistent from paper to paper. Indeed, even recent research papers use empirical calculations that are decades old and more than likely to contain oversimplifications or inaccuracies. Given the prodigious output of these stars even a slight initial inaccuracy will lead to a very different outcome for the star, as we shall see.

Yet, these inaccuracies don't come from apathy. The calculations required to determine how a rotating massive star loses mass through one mechanism or another are complex in the extreme and require enormous amounts of computing power. Consequently, most groups will attempt to trim the mathematics one way or another in order to obtain reliable results within a respectable amount of time. Many calculations take weeks on even the fastest computers, and it is hardly surprising that some mathematical fiddles are necessary to make the processes within the computer take somewhat less than astronomical times.

The largest problem is simply one of numbers. There are only a few dozen very massive stars known in the nearby universe. As each of these stars varies in its composition, its evolutionary state and the presence or absence of a companion, the role of mass loss through stellar winds remains unclear. Furthermore, how observed rates of mass loss are interpreted is also open to debate. Why? Well, consider an observation that an O-class star is shedding mass at a rate of one ten thousandth of the Sun's mass per annum. This figure is based on sensitive observations of wind velocity and density. However, these observations may assume a spherical outflow, where in fact magnetic fields or other factors cause more collimated or localized outflow. This would lead to overestimation of the rate of mass loss.

In addition, several observations suggest that very luminous stars have atmospheres that are so heavily polluted with material that instead of smoothly out-flowing gas, radiation pressure from the star drives material outward in some places, which then cools and falls back on the star in other locations in a manner akin to convection. Astrophysicist Stan Owocki suggested that this could lead to the atmosphere of a LBV resembling a Swiss cheese with the holes forming sites of upwelling and the areas of down-welling between them forming the walls of the cheese holes. Again, this kind of behavior could lead to easy misinterpretation resulting in the overestimation of the rate of mass loss. And these figures really matter. Mass loss rates for the most luminous stars vary from one hundredth to one hundred thousandth of a solar mass per year. This is a huge range.

Consider a star with an initial mass of 100 times that of the Sun, losing mass at a rate of one hundred thousandths of a solar mass per year. At this rate, by the time the star dies 5 million years later, it will have lost 50 solar masses, or half its mass. Increase this by a factor of 10, and the star is down to 5 solar masses and a likely death as a neutron star. Thus, getting these figures right across a range of stellar mass and composition is of fundamental importance to this area of stellar astrophysics. Observations, the foundation of empirical science, flounder here, as much of what is observed is open to interpretation. With limited numbers of very massive stars to observe, and with which to draw suitable conclusions, it is hardly surprising that there is little consensus at present.

The central issue emerging in the astrophysics of these massive stars revolves around what these stars actually look like when they age. O-class stars, with initial masses of over 40 times that of the Sun, are expected to expand modestly as they consume the available hydrogen in their cores. After a steady rise in luminosity throughout their main sequence lifetime, luminosity then remains fairly constant, as the star makes its exit to the right from the main sequence. However, as the surface expands and its temperature begins to fall the amount of mass lost from the star increases to perhaps a ten thousandth of a solar mass per year. The reasons for the increase in mass loss are discussed in more detail later, but even the simple act of extending the surface from the central store of gravity allows stellar winds and radiation to begin sculpting the stellar outer layers.

Stellar winds blow up to ten million or more times than seen at the surface of the Sun. Free electrons flowing within the atmosphere of the stars can intercept photons of radiation and accelerate. This has the effect of dragging the ionized gas as a whole along with them. If magnetic fields are present, and we have already seen some O-class stars have fields 1,000 times the strength of Earth's, and broadly equivalent to that seen in sunspots. These fields can collimate and help drive gas outward away from the star.

So, as we have touched on briefly already, just how much gas is lost from O-class stars and their immediate evolutionary successors?

A Problem with Wolves

What we do know is that older stars clusters, containing the most massive stars, also contain unusual stars called Wolf-Rayet (WR) stars, named after the discoverers of the first three members of this class, Charles Wolf and George Rayet. HD191765, HD192103 and HD192641 were discovered in the 1840s and were distinguished from the background O and B stars by their unusual spectra. A continuum punctuated by broad emission lines, indicative of hot, fast moving gas, characterized all three stars.

WR stars are clearly descended from O-class stars as evidenced by their mass, association and their unusual chemotypes. WR stars come in a few different flavors that are either hydrogen deficient or hydrogen free. O-class stars appear to evolve through an initial phase where the products of hydrogen fusion through the CN process become visible at the stellar surface. Progressive, severe mass loss brings nitrogen-14 up to the surface, as the hydrogen-rich envelope is driven off into space. This slowly changes the O-star to a WN (nitrogen) class star (Fig. 3.2).

As helium fusion sets in, carbon begins to emerge at the stellar surface, this time driven there by vigorous convection within

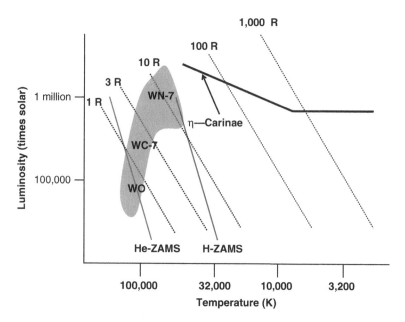

Fig. 3.2 The location of the Humphrey-Davison limit on the HR diagram (*solid, thick black line*) and relative position of Eta Carinae and some Wolf-Rayet stars (*blue shading*). Key: *WO* oxygen-rich WR stars, *WC* carbon-rich WR stars, *WN* nitrogen-rich stars, *R* refers to multiples of the solar radius, *ZAMS* Helium (He) or hydrogen (H) zero age main sequence for comparison. These lines represent the locations where stars begin fusion of these fuels for any starting mass

the now exposed helium core. The star becomes a WC Wolf-Rayet. Over time this may change again to oxygen-rich, and the star is then classed WO.

Although this sequence is fairly well understood in terms of nuclear reactions and stellar evolution, what remains unclear is the timescale, and whether all O-class stars pursue this route to eventual supernova. Moreover, there are a handful of very rare, highly luminous, hydrogen-rich stars called luminous blue variables (LBVs) that display periodic violent outbursts. These drive very vigorous mass loss, up to several solar masses at a time. Ten years ago many, perhaps the majority of, astrophysicists would have claimed all LBVs evolved into WR stars. The LBV stage represented a transitory phase between sedate mass loss on the main sequence and the final stages involving more advanced nuclear burning. Indeed, this is still the most prevalent view in the literature. However, it

Table 3.1 Various properties of the universe's most massive stars. Much of the data is inferred, although a few pieces, such as the masses of WR stars, are known from observation

Star type	Surface temperature (K)	Mass range (Solar masses)	Rate of mass loss
O-class Main Sequence	35–50,000	40–>150 (upper end of range unvertain)	10^{-5}–10^{-4} solar masses (likely lower at low metallicity)
LBV	12–35,000	>40 (upper end of range uncertain)	10^{-2} or more solar masses in outburst; 10^{-4} solar masses in quiescence
WR	40–80,000	5–48 from observation	10^{-5}–10^{-4} from observation

is becoming apparent that LBVs may represent the final phase in the life of the most massive stars. Some stars have been observed exploding, while LBVs and some of the most violent and brilliant explosions in the cosmos may be associated with these objects (Table 3.1).

The Humphrey-Davidson Limit

A zone of stellar avoidance exists in the Hertzsprung-Russell diagram, named after its discoverers, Roberta Humphrey and Kris Davidson. This zone is located above a stellar luminosity of roughly 300,000 times that of the Sun at the red end of the HR diagram. However, to the left of temperatures exceeding 4,000 K, the dividing line slopes gently upwards towards the main sequence, reaching luminosities of over five million times that of the Sun at its blue extremity.

The handful of stars that populate the upper edge of this domain lead precarious existences. All of these stars are blue. This is not a casual observation. Below this limit there are a few yellow and red stars, such as Rho Cassiopeia, but none exists above. The tiny smattering of blue stars, to the right of the Humphrey-Davidson limit, includes the Pistol Star, located in the Arches cluster near the galactic core. These few rarities are not expected to remain there for long, for reasons we shall discuss. Beneath the Humphrey-Davidson limit we encounter the first red supergiants as we descend and approach 100,000 times the luminosity of the Sun.

Why are there no million solar luminosity red supergiants? The answer lies again in opacity. At temperatures below 4,000 K all of the hydrogen in the outermost part of the envelope is present in the neutral state, but above 7,000 K it becomes ionized. As the hydrogen heats up and ionizes, its opacity increases. This traps radiation and drives expansion of the star, which, in turn, can lead to the shedding of mass in large pulses. This mechanism was initially proposed in 1962 by Robert Christy as a driving force in cyclical mass loss observed in many red giants and supergiants. Mechanistically, it is related to the pulsation mechanism in F-class stars, except that in these stars the pulses are driven by the ionization and recombination of helium at higher temperatures than seen in red giants and supergiants. However, Christy proposed that this mechanism would drive faster pulsations in red giant stars. LBVs display a more leisurely pulse that operates on decadal timescales. Therefore, Christy's mechanism may not fully extend the universe's brightest sparks.

At cool temperatures molecules begin to form along with dust grains. These effectively trap radiation and lead to the formation of powerful stellar winds. Radiation emerging from the stellar core drives matter away from the star as it is intercepted by these grains. This keeps the surface of these stars hot simply because the cooler gas is pushed aside into interstellar space. Stars above the Humphrey-Davidson limit simply act as skins in stellar winds. Once this skin has gone fully ionized, hydrogen is exposed at the stellar surface, and it becomes easier for the incumbent radiation to escape the star. This, in turn, increases the stability of the star overall.

Stars exiting the main sequence below this limit move to the right to cooler climes in the same manner as lower mass objects. However, stars above this limit encounter instability as a result of hydrogen recombination. It is expected that these stars may pulsate as they migrate towards the area of the HR diagram where hydrogen begins to cool sufficiently to recombine. These pulsations may drive mass loss, limiting the progress the star makes to the right of the main sequence. At the lower end of the mass scale (17–25 solar masses) pulsations are expected to be relatively weak but strengthen with initial stellar mass. Eventually, near 25 solar masses, the pulsations calculated by Sung-Shul Yoon and Matteo Cantiello are expected to be strong enough to drive energetic mass loss.

Mass loss rates detected for some of the more massive red supergiants, such as VY Canis Majoris, are in keeping with pulsation driven mechanisms. These could help drive stars away from the red toward the blue end of the HR diagram and generate the area called the "Yellow Void" at temperatures of 7,000–10,000 K. Above this, the Humphrey-Davidson limit is reached, and no stars are observed in the yellow or red portions of the HR diagram.

The stars rho Cassiopeia, and the less glamorously entitled IRC+10420, are interesting cases in point. Both stars lie just below the Humphrey-Davidson limit in the HR diagram and both have suffered periodic outbursts as well as episodes of dimming caused by the ejection of shells of dust-laden gas from its surface. Eruptions of rho Cassiopeia occur a few times per century and are associated with the ejection of shells of a sizable fraction of a solar mass.

Eventually, continuing mass loss will cause the star to travel backward toward hotter climes, becoming progressively bluer as much of the hydrogen-rich envelope is shed from its outer layers. In a few thousand to tens of thousands of years, rho Cassiopeia will emerge from its cocoon as a blue supergiant star, en route to its ultimate destruction as a supernova.

Luminous and Violent Blue Variables

Working higher up the main sequence, studies by Nathan Smith and Stan Owocki suggest that pulsations may shape the evolution of the most massive stars. At higher masses, there is a discrepancy between the numbers of WR stars and those that are expected to give rise to them. If O-class stars give rise to WR stars through a simple evolutionary process, then the mass loss rates observed for O-stars should be sufficient to produce these stars in the numbers that are observed. Quite simply, this is not the case. Mass loss rates are too low.

So, how do you square the round peg? The simplest answer is episodic, and violent, mass loss. LBVs eject mass in waves, detonations and pulses at significantly higher rates than either O- or WR-class stars. Indeed, such episodic mass loss is observed. In the Milky Way the LBV P-Cygni has repeatedly shed waves of material amounting to a tenth of a solar mass or so over the last few hundred years. There is a similar story for AG Carinae, but with somewhat higher masses (a few solar masses) involved in the ejections.

Periodic outbursts on LBVs are well known and earn these stars the title S Doradus variables, after the progenitor in the class. Lesser explosive events of LBV stars during their quiescent state take the normally blue star to cooler temperatures, but with roughly constant luminosity. This feat is achieved by the lowering of temperatures accompanied by an expansion of the envelope as material is driven outward. These eruptions appear to match the scenario suggested for lower mass stars such as rho Cassiopeia, with eruptions driven by ionization of hydrogen or, at higher temperatures, helium just beneath the photosphere of the star.

However, although less common a few LBVs display more violent and less frequent explosions. One such star is Eta Carinae. This star is renowned for its massive outburst in 1843 where 12–20 solar masses of material was violently ejected. The question is, what mechanism drove the instability that resulted in this extensive, episodic wave of mass loss? Mass loss from the 1843 eruption has repeatedly been revised upward, and recent observations by Nathan Smith indicated that in addition to the closely observed Homunculus Nebula, a bipolar feature surrounding the star and its unseen, massive companion, an additional waist was present around the midriff of the Homunculus. This waist contained material expanding radially outward at 3,500–6,000 km/s. True supernovae may have expansion velocities only marginally faster than this. Typical LBV wind speeds are closer to 100 km/s; thus, the already extraordinary eruption of 1843 just became even more interesting.

The Homunculus has an expansion speed closely matching the escape velocity of the star, indicating that although it probably began its journey from the parent star in the 1840s eruption its expansion is largely driven by stellar winds from the star since the outburst. However, the faster moving waist must have a different origin intrinsic to the mechanism of the eruption itself. Moreover, Smith's work increased the total mass lost in this single violent episode. Earlier estimates put the star at a few solar masses at most. Smith's work increased this to nearly 20 solar masses – a truly enormous explosion, particularly since the star remained intact after this cataclysm.

In a nearby galaxy, NGC 3432, a similar event to Eta Carinae's outburst was seen unfolding in the year 2000 – Supernova 2000ch. Although the spectra were clearly inconsistent with a true supernova, material was seen moving away from the central star at over 1,550 km/s, and the luminosity peaked at an absolute magnitude of −12.7, or tens of millions of times brighter than the Sun. The central star clearly survived this outburst, but its extreme luminosity – at least for a non-lethal event – suggested that a considerable amount of material was ejected. Together, these especially violent explosions suggest mechanisms exist within the most massive stars that drive violent eruptions. In turn, these violent eruptions drive down the mass of the star immediately prior to the death of the star as a supernova. What mechanisms exist in the astrophysicist's toolbox that might explain these eruptions?

One possible answer is simple instability in energy transport in the envelope of the star. We have already encountered this in this chapter, and it may well explain the presence of the Humphrey-Davidson limit in its entirety. Very high luminosities might drive convection-driven eruptions or pulsation-driven shell ejections. However, these sorts of mechanisms seem insufficient to explain the truly massive outbursts of Eta Carinae and SN 2000ch. Instabilities in energy generation are another possibility. LBVs should be at the evolutionary stage where energy generation by the CN cycle is winding down as core hydrogen levels fall. This is evident in the abundant helium and nitrogen found in the winds of these stars.

Core helium ignition is thought to be a fairly passive process in massive stars, with a gentle ignition once the temperature exceeds 100 million degrees. However, the core of a very massive star may reach over 200 million Kelvin, and it may be that helium fusion ignites or burns in a more unstable manner. There is a very steep dependence on the rate of burning of helium with temperature. Thermal excursions could occur, where the temperature rises throughout the core as a result of pulsations in the overlying envelope. Whole core, or star, pulsations might then drive unstable burning. Perhaps the core begins to pulsate, and like the famous Tacoma Bridge, these pulsations build harmonically and simply get out of hand. At some critical frequency the core overheats. This drives a wave of nuclear burning that in turn directs a violent pulsation. This in turn sheds a violent blast of material from the envelope. Each ejection should then change the dynamic and harmonic frequency of the star, altering any future pulses and outbursts.

Possibly a simple instability in the rate of hydrogen burning in the shell that surrounds the core is the source of instability. In some evolved stars, once the shell in which hydrogen fuses comes close to the stellar surface nuclear burning becomes intermittent. This is particularly true if the depth of envelope above the burning shell is narrow. Alternatively, more advanced fuels may undergo violent excursions in the rate at which they burn. It is already known that oxygen burning can be unstable in particularly massive stars, when the temperature exceeds a billion Kelvin or so. Could something like this lie behind the larger eruptions of LBVs?

Instability in oxygen burning occurs through the pair-instability mechanism, whereby high temperatures lead to the creation of electron-positron pairs. Creation of these pairs removes internal support for the star leading to core collapse, violent heating and a subsequent explosion. In theory, stars with masses comparable to Eta Carinae may undergo these sorts of explosions but survive. However, pair-instability is expected to arise during oxygen burning, which according to theory shouldn't occur more than a year before the death of the star. Neon burning, occurring immediately prior to ignition of oxygen, is similarly brief, leaving instability in helium or carbon burning as the only alternative nuclear mechanisms for these pulsations.

In their 1994 paper, Roberta Davidson and Kris Humphreys suggested that the source of instability probably lay with the mechanics of energy transfer in the envelope as the star approaches the Eddington limit at which radiation pressure overwhelms the effect of gravity. Alternatively, it was suggested that changes in the structure of the star as it evolves off the main sequence might destabilize the star. At the time – and indeed until very recently, these suggestions would appear to be the most likely mechanism.

This was believed given what was known about the likely stability problems these stars would encounter as a consequence of their extraordinary energy output.

However, given that LBVs appear to show fractious instability in the years or decades prior to core collapse, it now appears that simple mechanisms involving structural changes to the envelope or energy transfer are unlikely to lie behind these near catastrophes. Something far more fundamental must lie behind (at least) a proportion of these outbursts; they occur too close to stellar death to be linked to the passage of the star from the stellar main sequence.

At present we can only speculate as to the underlying mechanism, or mechanisms, behind these extreme outbursts. Improvements in our understanding of the basic physical processes will be required.

Evolutionary Paths of the Most Massive Stars

Evolution of the most massive stars can be understood as a race between the evolution and death of the stellar core and the loss of the stellar envelope and outer core through stellar winds. In some regards this race parallels the late evolution of stars such as the Sun, but unlike this, the process of mass loss involves the core of the star directly.

In low and intermediate mass stars in the latest stages of evolution, the so-called asymptotic giant branch (AGB) stage, the stellar core is growing slowly through the unstable burning of hydrogen and helium at its surface. Core growth is terminated by the loss of the bulk of the stellar envelope through stellar winds, a process that last a few hundred thousand years at most. Stellar winds remove approximately one hundred thousandth of a solar mass per year. Conversely, in very massive stars the loss of mass runs at a similar rate throughout the main sequence and onward throughout the WR phase. However, the 40,000–100,000 year gap between these stages shows an increase in mass loss to perhaps a hundred or a thousand times this rate. As we have seen, mass loss at this rate is achieved through violent explosive events that shed several solar masses at a time, eliminating the thick hydrogen envelope in its entirety.

This evolutionary scenario plays out on the HR diagram as follows. The star initially takes a path to the right from the hydrogen main sequence (H-ZAMS) towards the Humphrey-Davidson limit, where mass loss, through various possible mechanisms, results in the removal of the stellar envelope and the uncovering of the helium core. As mass is shed the star loops backwards toward higher temperatures, crossing the main sequence toward a hot, blue tract called the helium-burning main sequence (He-ZAMS). With surface temperatures perhaps approaching 100,000 degrees Kelvin, helium is consumed within the core of the star, while stellar winds continue to whittle down the star's overall mass. As the mass declines, perhaps to as little as a few solar masses, the overall luminosity declines, and the star descends the helium main sequence. Here helium burning continues, and as the outer portion of the exposed core becomes polluted by heavier elements, the star may re-expand and cool, crossing back to the right of the hydrogen main sequence. Shortly thereafter, the core collapses and the star detonates as a supernova, terminating further evolution.

In the HR diagram, this is the standard model for the most massive stars. However, it may not be true for all such stars. The key issue is mass loss. If the star retains a thick hydrogen layer, it will remain on the cooler side of the main sequence on the HR diagram after core hydrogen burning ceases. Evolution of the stellar envelope may proceed at a slower pace than the core, and core collapse may ensue before the hydrogen layer is removed. In this scenario, how might evolution on the HR diagram play out?

Staying to the left of the Humphrey-Davidson limit places the star in a difficult evolutionary bind. Evolution in the core is trying to drive it to the right, while limits to energy transport are trying to drive it to the left, onto the helium burning main sequence. Consequently, instability within the star drives eruptions from the stellar surface, and the star undergoes waves of mass loss. Such stars, Eta Carinae included, periodically brighten and fade and are known as supernova imposters. Luminosity at their peak approaches millions of times that of the Sun, and material is violently ejected from their surfaces.

Aside from Eta Carinae and SN 2000ch, several more examples of these supernova "imposters" are known. The Type II explosions SN 2002kg and SN 2003gm were characterized by Justyn Maund and others (University of Copenhagen). Although these explosions were initially considered as true core-collapse events, their low peak luminosities and sluggish blast waves revealed a different story.

SN 2002kg had a peak absolute magnitude of -9.6, easily within the range of LBV outbursts. The spectrum also revealed another key feature seen in LBVs: Iron-II (Fe(II)) emission. Taken together, this "supernova" was clearly a so-called imposter, a non-lethal stellar explosion masquerading as a supernova. Striking with somewhat greater oomph, SN 2003gm had a peak absolute magnitude of -14, but the progenitor appeared to be a yellow hypergiant, similar to rho Cassiopeia. It remains unclear if SN 2003gm was a true supernova, but simply a weak, low luminosity event, or an outburst similar to Eta Carinae. The luminosity was rather high, but not unreasonably so, for an imposter. Further observations of the progenitor's location are required to determine if the star is indeed deceased or simply hiding behind a prematurely fitted death mask.

Type IIn Supernovae

As we enter the realm of the most massive stars, it is worth describing the possible types of supernovae these beasts may become when they die. If the star manages to retain its hydrogen envelope the supernova generated will be Type II. However, as we have seen, the most massive stars will actively be shedding this layer as they approach their demise. Consequently, the star is likely to be surrounded by extensive debris of its own creation. When the star subsequently explodes the supernova blast wave will sweep this material up, and the ensuing light curve will be strongly attenuated by the effects of this collision. The explosions are thus unique and have the designation Type IIn, reflecting the effects of the collision between the blast wave and the surrounding debris.

Type IIn supernovae have a number of distinguishing features that set them apart from other, more typical Type II-P events. As we have already seen Type II-P explosions display broad, so-called

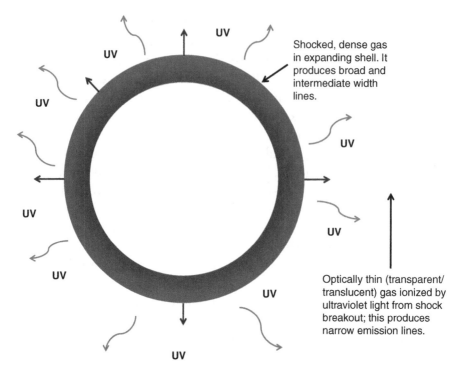

FIG. 3.3 The formation of narrow and broad emission lines by shockwave interaction with circumstellar gas in Type IIn supernovae. The faster the gas is moving, the broader the lines produced

P-Cygni, absorption lines produced by the expanding shell of debris from the exploded star. Paired blue- and red-shifted absorption and emission are indicative of hot hydrogen-rich material moving outward from the center of the blast at high speeds. The blue-shifted emission lines are created by matter moving towards the observer, and the red-shifted lines by material moving away from the observer at the same speed but in the opposite direction.

In these explosions the decline in luminosity is clearly powered by radioactive decay. By contrast, the light of Type IIn explosions is bluer than that of Type II-P explosions. The supernovae spectra usually lack the characteristic broad P-Cygni lines, and significantly the explosions are on average brighter, have longer rise times to maximum light and show prolonged decays in their light curve. Importantly early spectra contain narrow emission lines that are unique to these events. It is these narrow lines that give the explosions the IIn designation (Fig. 3.3).

Explaining Type IIn Properties

The principle characteristics of Type IIn explosions can be understood in terms of collisions. In a Type II-P explosion the energy powering the emission predominantly comes from the decay of radioactive cobalt and nickel-56. The details of these explosions, such as the plateau in the light curves, are produced by the recombination of hydrogen ions and electrons to form neutral hydrogen. Recombination occurs at temperatures of 4,000–6,000 K and leads to a period of near constant luminosity in the light emitted by the supernova.

Once the plateau phase of the light curve has ended, the light declines in a manner dictated by the radioactive decay of nickel-56. By contrast, Type IIn supernovae are characterized by narrow emission lines indicative of flash ionization of circumstellar matter, and more significantly the presence of intermediate and broad emission lines that bear witness to the violent collision of the supernova shockwave with slower moving material surrounding the star. Spectra reveal the presence of a slow-moving mass of material expanding at 100–200 km/s, through which the 10,000–15,000 km/s supernova blast is plowing (Fig. 3.4).

Few Type II-P supernovae show much evidence of interaction of the blast wave with surrounding matter. A clear and interesting exception is SN 1987A. Just over a decade after Sanduleak −69 202A exploded, the blast wave rammed into a clumpy ring

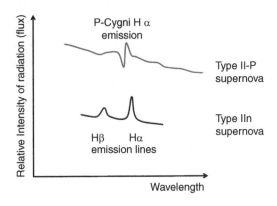

Fig. 3.4 The classic P-Cygni line of a Type II-P supernova versus the narrow emission lines of a Type IIn event

of hydrogen-rich matter shed millennia earlier by its parent star. Although the material was spread too far from the supernova to generate the narrow lines seen in Type IIn events, the violent interaction has resulted in a brightening of the supernova remnant seen at visible, X-ray and radio wavelengths. A more recent supernova, SN 1996r, showed a similar brightening in 2001, coincident with the blast wave from the supernova slamming into a shell of material ejected centuries earlier by the parent star. Indeed we can hypothesize that Type II events are differentiated into Type II-P and Type IIn depending on when the phase of greatest mass loss occurred relative to the terminating supernova.

In Type IIn explosions the collision of the blast wave with surrounding material explains various other features of the display. As the rapidly expanding material plows into the material around the star (circumstellar material or CSM for short), shock heating generates the profuse X-rays and radio waves that are often characteristic of these explosions. The efficient conversion of the kinetic energy of the blast wave through the collision generates the visible light. The light is also typically bluer in these explosions as a result of the higher energies involved, hence other observations regarding the color of these blasts.

If the dense material is spread over considerable distances from the star, the collision may be prolonged, subsequently powering a lengthier display than is seen in Type II-P events. A few examples of Type IIn explosions show very prolonged rises to maximum light (more than 50 days), and many show relatively sluggish declines in absolute magnitude. These differences appear to be caused by the lengthier battle between the outgoing shock and slower moving, denser material expanding away from the progenitor star. In contrast, the decline in the visible output of Type II-P explosions largely matches the decay of radioisotopes such as cobalt-56 and titanium-44.

Catching the Wave

Supernova PTF 09uj presented an interesting addition to this picture. Discovered on June 23, 2009, by the Palomar Transient Factory (PTF), there was a serendipitous and nearly simultaneous observation made by the GALEX ultraviolet observatory. GALEX spotted a rise in ultraviolet emission from the region of the sky in which the supernova erupted in the days prior to the visible outburst. The UV emission rapidly peaked at a substantial absolute magnitude of −19.5. The rise in UV was interpreted by Eran Ofek and co-workers as the shockwave breaking out from the "surface" of a shell of dense circumstellar material that surrounded the supernova.

This interesting observation pointed to a phase of intense mass loss that occurred only a few years prior to the supernova. Mass loss was in the region of a tenth of a solar mass per year – far in excess of the levels normally observed in stellar winds and suggestive of some form of outburst having happened in a short space of time before the star exploded. In this scenario, the star vented a large amount of mass, then core collapse triggered a supernova. The blast wave plowed through the circumstellar material, generating the characteristic emission lines of a Type IIn event before punching outward into the surrounding interstellar medium. The emergence of the shock generated the strong UV pulse detected by GALEX.

Type IIn explosions therefore represent stars *in flagranté delicto* – caught in the act of shedding their skins immediately before their death. Less massive stars undergo more modest, if not less prolonged, episodes of mass loss as they evolve into red supergiants. Stellar winds are able to keep pace with the overall speed of mass loss and keep the shed material relatively well dispersed in the vicinity of the star. The behavior of less massive stars is reminiscent of scenes from *The Great Escape* where Steve McQueen was patiently dispersing darker tunnel soil in among the lighter soil of the field. The idea was that the prospective escapees would remain undiscovered by their hosts. In contrast, the progenitors of Type IIn explosions have no such qualms and pump vast quantities of their bulk into the surrounding interstellar medium. These stars simply can't lose weight quickly enough before they die.

So, why do these stars behave so differently? This isn't entirely clear, as, with a few exceptions, not all progenitors of Type IIn explosions are identifiable in archived images. Those progenitors that are will be the very brightest stars visible across intergalactic space. Therefore, there is a strong selective bias in what is observed, and the results may not, therefore, be representative of Type IIn supernovae progenitors overall. What can be said at present is that the progenitors of the brightest Type IIn explosions appear to be particularly massive stars, with many apparently exploding while they are luminous blue variables. Notable examples include or are inferred to be the progenitors of SN 2005gl, SN 2006jc, SN 2006gy, SN 1994W and the progenitor of SN 2008iy. That said, the only one that is clearly identified in archival images is the progenitor of SN 2005gl. The status of the others is based on inferences from the explosion characteristics, or in the case of SN 2006jc, a coincident LBV-like eruption at the site of the supernova 2 years prior to the explosion. Undoubtedly, as more observations are made, the progenitors of more and more supernovae will be identified, and sooner or later an LBV star will get caught in the act.

Type IIn events also carry enormous diagnostic insight. As the explosions drive into and illuminate the circumstellar material, they afford astronomers glimpses of the star in the decades running up to decimation. Light from the supernovae may illuminate surrounding matter directly – and this is suggested in the early spectra of SN 2006gy. Collisional excitation of the circumstellar material by the outgoing shock can also reveal the chemistry of the gases, which in turn opens windows on nucleosynthesis and mass loss in these highly evolved stars. Future observations of Type IIn explosions are expected to expand our understanding of the stars that give rise to some of the brightest supernovae.

With this in mind, it is now worth considering which sorts of stars give rise to these supernovae. Two generic possibilities exist: low mass, so-called super-asymptotic giant branch (hereafter abbreviated to SAGB) stars, and very massive LBV-like stars. In Chap. 7 we will examine the electron-capture route through which the SAGB stars are believed, in some instances, to die. However, it isn't yet clear whether any so-called electron-capture supernovae have actually been observed. Many of the observed dim or peculiar Type II explosions can be ascribed to fallback onto a developing black hole (Chap. 5), while some odd chemically peculiar Type I supernovae, such as SN 2005cs or SN 2005E, seem to be linked to white dwarf stars – not massive stars. The faint SN 2008S considered in Chap. 7 may not even be a true supernova – and this may also be true for other intrinsically faint Type

IIn events. Instead they may be supernova imposters, such as SN 2002kg. The reminder of this chapter will be devoted to a handful of the brightest, longest or most informative Type IIn explosions. In Chap. 8 we will examine two of the models that might explain the brightest Type IIn events.

From Imposter to Supernova

So, some very massive stars may undergo explosions of sufficient luminosity to be mistaken for supernovae. These outbursts occur during the LBV phase. But what if the star has evolved somewhat further before core collapse – but not so far that it appears as a WR star? Enter SN 2006jc – a peculiar Type Ic explosion that seemingly occurred within a dense shell of helium. The spectrum of this explosion had a few antecedents but was particularly unusual. Although not spectacular the supernova was relatively bright, peaking in excess of the majority of Type Ic explosions. Of greater interest was the unusual spectrum that revealed prominent narrow emission lines of helium (He I). The authors of the 2008 *Nature* paper suggest calling this type of supernova Type Ibn to signify helium in the spectrum but also that the helium lines are narrow in a manner similar to Type IIn hydrogen-rich events. These supernovae gain their narrow lines from collisions, and the luminosity is thus powered by both collision as well as the decay of radioactivity or photospheric expansion.

The supernova occurred at a spatially coincident point, with an earlier supernova imposter. Japanese amateur Koichi Itagaki identified a brief but prominent explosion at the same location in 2004. This first explosion peaked at an absolute magnitude of −14.1, bright but not unreasonably so for supernova imposters. This implied that during this LBV-like outburst the star ejected its helium-rich outer core and then continued onto core collapse 2 years later. In this short window, the star became a WC or WO-class Wolf-Rayet object, which then died as a Type Ic supernova. The supernova blast wave then collided with the helium-rich shell ejected 2 years earlier. Although LBVs are hydrogen-rich, it is not unreasonable to assume they could evolve during an unstable intermediate phase before becoming WR stars. In this instance,

SN 2006jc simply occurred before significantly more evolution of the stellar envelope could occur and the WR stage was cut short.

High resolution spectra reveal the presence of more distant shells, implying that the progenitor underwent several rounds of violent mass loss prior to core collapse. Is this a fate awaiting other LBVs? These massive, highly luminous stars are so rare that productive empirical modeling of their evolution based on observation is effectively ruled out. But we are left with the possibility that LBVs may meet a fate quite unlike WR stars.

More recently, Eran Ofek observed the Type IIn Supernova SN 2010mc (PTF 10tel) and noticed that this explosion must have been preceded by an outburst 40 days prior to the big blast. Around 2 weeks after the supernova initially peaked at an absolute magnitude of −15, there was a second surge in luminosity to −18.4. Spectroscopic evidence indicated that this second surge was caused by the collision of the 10,000 kilometer per second shockwave with a slower moving (2,000 kilometer per second) outer shell. The distance and relative speeds gave the time of the first eruption. Ofek suggested that this pre-supernova eruption was best explained by a change in the rate of nuclear reactions within the core of the former 50 solar mass star. Perhaps coincidental with the ignition of one fuel type, the change in the pace of the nuclear fires generated first violent convection within the core. This, in turn, spawned a wave within the surrounding envelope of the star, ultimately blowing 0.02 solar masses (one fiftieth of the Sun's mass) of material out into space. Although the model is unconfirmed, it goes some way to explaining why some particularly massive stars appear to become unstable shortly before death. Did the progenitor of SN 2006jc undergo a similar nuclear-driven outburst? At present the math will have to catch up with observations.

Returning to the fate of massive stars, why is it that some massive stars evolve to WR stars while others stay as LBVs? Perhaps instability on helium ignition is the issue or perhaps the answer lies in the rate of rotation. Massive stars that rotate quickly could lose mass more effectively through stellar winds from their equatorial regions. Slower rotating stars may cling on to their mass. However, observations of WR stars don't show any preference for mass loss around their waists. Mass is seemingly lost across the whole surface of the star. So, if this is a possibility, it can only be for some WR stars. Metallicity, or stellar chemotype, may play a role; however, once again, the relationship between mass loss and chemotype is very messy and poorly constrained at present.

Direct Detonation of LBVs

If SN 2006jc truly represents the death of a star transitioning from LBV to WR, then what if the star explodes even earlier in the evolution of the envelope? A few interesting supernovae have emerged in recent years that appear to represent such a demise. SN 2005gl is a case in point. A Type IIn explosion, although not unreasonably bright, displayed narrow hydrogen emission lines in its spectrum indicative of a collision between the supernova blast wave and surrounding shells of hydrogen-rich material. The progenitor star, NGC266-LBV 1, was identified as an LBV, making a clear case for the revision of stellar evolutionary models – LBVs can explode. Although SN 2006jc may be regarded as a peculiar, marginal case of an LBV-like progenitor that was identified by Douglas Leonard and Avishay Gal Yam, the star was clearly an LBV with a luminosity roughly ten times that of Sanduleak-69 202, the progenitor of SN 1987A. Thus we are confronted by the scenario whereby LBVs can and do explode without recourse to further evolution through a Wolf-Rayet stage.

Early spectra reveal that the supernova was surrounded by a shell or shells of material that had been shed at a rate of roughly 0.03 solar masses per year. This mass is similar to many LBVs, such as P Cygni, although it lies well below the violence of the extreme end-member Eta Carinae. This hydrogen-rich material was expanding at a little over 1,500 km/s – speeds typical of LBV eruptions but considerably faster than typical non-eruptive winds from LBVs or their less showy red supergiant cousins. By day 58 the spectra revealed that the shockwave had already bulldozed its way through the circumstellar material and was now freely expanding at 10,000 km/s. Light was now coming directly from the supernova ejecta, not from the interaction of the expanding shock with the surrounding stellar debris. The progenitor LBV, itself, was relatively metal rich.

This poses another problem for stellar evolutionary modeling. If we assume that metallicity increases mass loss, why did the progenitor of SN 2005gl explode as an LBV and not a WR star? Just how did it hang on to its mass? We might then infer that the progenitor was exceptionally massive, and its large initial mass allowed it to retain sufficient hydrogen to appear as an LBV at death. However, if we descend this path we hit another snag. Very massive stars are predicted to encounter pair-instability and die through different mechanisms that are developed in Chap. 7. However, there is no evidence that SN 2005gl underwent any sort of nuclear instability. Something fundamental has to give. The evolutionary models we have – at least the most popular ones – do not match observations. Problems of initial stellar mass, mass loss, metallicity and even the basic mechanisms of stellar death are on the ropes at this point. It cannot be stressed enough that these problems are fundamental, and clear solutions are not yet in sight.

Leaving these vexing issues aside for now, we will examine a few more supernovae associated with luminous blue variable stars, if only to underscore the problems current models of stellar evolution face. If the Type IIn Supernova SN 2005gl was a little underwhelming in terms of sheer output, a few more Type IIn supernovae are known that show many more intriguing features that illustrate just how the processes that degrade LBVs can produce remarkably powerful or long-lasting explosions. Without seeking to exhaustively discuss them all a few notable examples will be discussed here in preparation for a more thorough dissection of some of these supernovae in later chapters. As we approach these more unusual supernovae we will be underscoring the emergence of new frontiers in stellar astrophysics.

Slow Blow: The Case of Supernova SN 2008iy

Typical core-collapse supernovae show rise times (the time from explosion to maximum visible light) in the region of 20–30 days. This is the time it takes the photosphere of the supernova to grow to maximum and energy from radioactive decay to diffuse outward to become visible near the supernova photosphere. However, supernovae powered by collisions may show more extended emission that takes longer to rise to maximum. A handful of supernovae take 50 or so days to rise to peak luminosity, but one supernova, SN 2008iy, holds the record. From detonation to peak light, this event took place over 400 days, and at 800 days the explosion was still visible. The peak luminosity was high, but not unreasonably so at an absolute magnitude of −19.1. However, over the first 700 days the supernova emitted 2×10^{43} J of energy in visible light, or roughly 10 times that emitted by the average Type II-P supernova. Where did this extra energy come from?

Let's look back to the figures for a core-collapse Type II-P supernova. A typical Type II-P supernova releases a total energy of around 10^{46} J. Of this 99 % is emitted in the form of uncharismatic neutrinos, with 1 % in other forms. Of the seemingly trivial 1 %, 10^{44} J is in the form of kinetic energy and 10^{42} J is in the form of radiant light and other types of electromagnetic energy. Radiated light is clearly not the power source, but the conversion of only 10 % of the kinetic energy to light is easily sufficient to power these extended and overly luminous displays. The presence of narrow Balmer emission lines in the spectrum of SN 20008iy indicated flash ionization of reasonably dense, hydrogen-rich circumstellar material. And finally, the presence of intermediate width emission from hydrogen clearly demonstrated the effect of a collision between the supernova shockwave and surrounding dense material (Fig. 3.5).

Later spectra suggest that the surrounding material was expanding away from the progenitor star over the century or so preceding the explosion and that the speed of expansion was typical of luminous blue variable eruptions (greater than 100 km/s). The interesting feature of this wind is that it appeared to be strewn with clumps of denser, optically dense gas and possibly dust evident as those intermediate width emission lines. Spectra indicated that these lumps had an expansion speed from the star of around 1,650 km/s, considerably slower than the 4,500–5,000 km/s speed of the blast wave. The rise in overall emission suggested that the density of clumps increased with distance from the parent star. This, in turn, suggested that some form of eruption, typical of an LBV, occurred 55–100 years prior to the supernova. However, Adam Miller noted that the clumpy CSM could be distributed in a bipolar structure around the progenitor much like is observed around Eta Carinae or SN 1987A. The clumps then form the walls

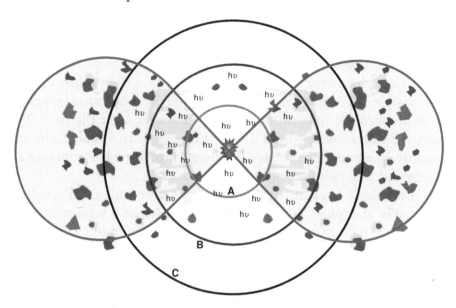

FIG. 3.5 A model that explains the properties of the exceptional SN 2008iy. The supernova detonates, and the UV flash illuminates the inner layers of the ejecta. Most is absorbed further out. This generates the narrow emission lines in the spectrum. The blast wave moves outward over time (*A*, *B* and *C*), intercepting more and more clumps of dense circumstellar material that increase in density with distance from the supernova. The circumstellar material may either be arranged in a shell or in bipolar lobes as shown

of the structure through which a more radially symmetrical blast wave is advancing.

What else was gleaned from the spectra of the explosion? Mixed in with the signal of the collision are fainter narrow P-Cygni lines indicating that the clumps are embedded in optically thin material that was flash ionized by the ultraviolet light from the shockwave breaking out of the progenitor star. The presence of clumpy material embedded within a finer matrix of optically transparent gas is also noted for the ultra-luminous supernovae SN 2006gy and SN 2006tf, which are described later, clearly linking these two kinds of events.

Finally, SN 2008iy was a strong source of X-rays. These are the expected products of the violent collision of a shockwave with hot gas. The shockwave violently heats and ionizes the gas at the boundary of the shock, and the electrons that are liberated release X-rays as they careen around the ions in the gas or are recaptured.

Strong X-ray emission is a clear diagnostic feature of collisions and is typical of Type IIn supernovae or the later stages of Type II-P explosions when much of the surrounding interstellar and circumstellar material has been swept up by the expanding shock wave.

Many interesting and informative comparisons can be made with other Type IIn explosions that illuminate the past histories of the parent stars. SN 1994W showed strong additional luminosity in early times, but after day 100 this dropped off precipitously. This indicated that the progenitor star underwent an LBV-like episode of extreme mass loss only a few years prior to the supernova, and that it took less than 100 days for the blast wave to punch through the material and out into the surrounding, low density, circumstellar material. SN 2006gy showed prolonged extreme luminosity, making it the third most luminous supernova prior to 2010. Here, the collision was prolonged and violent, indicating an episode of extreme mass loss, akin to Eta Carinae, which occurred shortly before the star exploded. Many other but not all super-luminous supernovae, or VLSNE, as they are known (very luminous supernovae) show strong similarities to SN 2006gy, indicative of violent interactions between the supernova shockwave and dense circumstellar material.

Are there any other potential sources of power for this display? Potentially very high abundances of radioactive nickel – several solar masses – could power a display lasting over 2 years. However, the influence of this power source would be apparent in the light curve of the explosion. A large supply of nickel-56 would produce a fast rise time, not one extended over 400 days. Similarly, the decay of this prodigious amount of nuclear fuel would be visible in the declining light curve once the explosion had passed its peak intensity. However, SN 2008iy shows little evidence of additional power from radioactive decay. Therefore, the inescapable conclusion was that this elegantly prolonged event was a consequence of an extended collision lasting over a year. Remarkable.

The Impact of Collisions (No Pun Intended!)

A handful of ultra-bright supernovae have been discovered in recent years that push the boundaries of how brilliant an explosion can be. SN 2006tf was discovered on the 12th of December, just before the close of 2006. The year 2006 was particularly productive in terms of its yield of ultra-luminous and informative explosions. As well as SN 2006tf, we had SN 2006jc, SN 2006gy and the enigmatic SCP 06F6 (Chap. 10). Part of the reason for this flurry of activity – barring sheer luck – was the growing impact of automated computer searches. SN 2006tf was discovered by the Texas Supernova Search as part of the automated ROTSE Supernova Verification Program, RSVP for short. Later photometry was obtained using another automated program, KAIT, the Katzman Automatic Imaging Telescope, run by Alexi Filippenko. Once again the supernova bore the clear signature of a collision between copious shells of matter and was classified as Type IIn. At the time the description was reported the explosion ranked third in terms of peak luminosity at an absolute magnitude of –20.7, making it considerably more luminous than SN 2006jc (and subsequently the peculiar supernova SN 2008iy). Only SN 2006gy and SN 2005ap exceeded the peak luminosity at the time of publication.

SN 2006tf was located in a dwarf galaxy – something of a common feature in many of the peculiar supernovae that populate this book. This may be more than coincidence. Dwarf galaxies tend to have lower metallicity than larger galaxies such as the Milky Way, and this fact may be related to the types of stars that generate unusual core-collapse explosions.

Estimates of the total radiated energy exceeded 7×10^{43} J, far in excess of the total budget of radiant energy of a typical Type II-P supernova. The total radiated output pushed the envelope of how much visible radiation could be drawn from the kinetic energy present in the blast wave. The typical figures hover around 10^{44} J; therefore, the amount of energy extracted from this material was nearly 100 %, if the initial energy in the explosion was typical of a core-collapse event. The question arose, how can you power a supernova this brilliant? Is it purely the product of a grand collision? Spectra taken early in the explosion revealed that the shockwave from the supernova had impacted a dense and optically thick shell of hydrogen-rich material. Almost all of the blast wave's kinetic energy was converted into visible light by the collision, and the blast wave decelerated to 2,000 km/s. However, was the initial explosion also more energetic at its outset? Perhaps the explosion was abnormal, driven by a jet or by instability within the core of the star? Initially, this wasn't clear.

However, later spectra began to reveal more of the inner workings of the explosion. Later spectra of SN 2006tf indicated that within a few weeks of detonation, the collision between the blast wave and the dense circumstellar material (CSM) was over, and now the shock was traversing less dense and optically thin, but still clumpy, gas similar to that surrounding the extraordinary supernova SN 2008iy, described above.

The 50–100 day rise time in the luminosity of the explosion suggested that the collision commenced at a radius of 180 astronomical units (A. U.) from the parent star, or 6 times the distance between the Sun and Neptune. This collision continued to distances exceeding 300 A. U., indicating that the blast wave was traversing a thick shell of material ejected from the star, perhaps only 5 years prior to the supernova. The presence of a single thick shell also indicated that the event that ejected the material was sudden and violent, perhaps akin to the giant eruption of the LBV Eta Carinae in 1843 or to the violent ejection of a shell in a pulsational pair instability event.

Pair instability will be described in greater detail in later chapters and was briefly mentioned earlier. However, in outline, oxygen burning may become highly unstable in very massive stars, leading to the ejection of the bulk of the hydrogen envelope in a single, violent pulse. The pulse carries a very similar energy to a supernova and may be observed as such – a supernova imposter. Stan Woosley's work suggested that pulsational pair instability (PPI) would occur a handful of years prior to the collapse of the core in some massive stars – those born with an initial masses in the region of 95–130 solar masses. Woosley and others published their paper in the journal *Nature*, suggesting that the violent explosion of SN 2006gy (Chap. 6) was a pulsational event. However, it turns out that SN 2006tf fits the bill for this kind of event even better than SN 2006gy.

The presence of a dense shell, expanding into a clumpy wind, is similar to the scenario proposed for an LBV that then encounters pair-instability. During the earlier phases of declining health, the LBV drove blobs of matter outwards in a dense, clumpy wind moving at speeds typical of LBVs (190 km/s). These amounted to a few solar masses, ejected during the last few decades of the twentieth century. Later, around the year 2001, the core became unstable through pair instability. This resulted in the ejection of an 18 solar mass shell of material in an unobserved super-outburst. Had this been witnessed it would have been considered a supernova imposter. The star survived this brief cataclysm, later to be torn apart by a final core collapse in 2006. Stan Woosley's best fitting model of SN 2006tf involves a 25 solar mass shell ejected by a 110 solar mass progenitor 5 years prior to core collapse, uncannily close to observation. The implication is that stars that retain around 100 solar masses in the decades prior to annihilation are subject to pair instability and die through this mechanism.

The key to understanding these explosions will lie in the detection of their earlier violent outbursts, characterizing the stars that give rise to them and then working forward to their final demise. However, this is a task somewhat harder to actually achieve than it is to put into words. After all we live in a universe bursting with trillions of stars. How many prying eyes do we have?

What of the earlier life of SN 2006tf progenitor? Unlike many earlier observed Type IIn supernovae, the late decline in luminosity of SN 2006tf was a gradual event, indicating that the wind ejected by the progenitor was extensive and had been blowing for hundreds of years prior to the expulsion of the shell. By contrast, SN 1994W showed a sharp, linear decline in luminosity after day 100, indicating that the shockwave had traveled through the dense material around the star. Thus the progenitor of this explosion had only begun a vigorous phase of mass loss for a comparatively short period prior to the star's destruction. Taken together it paints a picture of great variety within the fold of Type IIn supernovae, which in turn reveals a diversity of progenitor histories. These explosions simply can't all be lumped together. Apparently many roads lead to Rome.

Despite this neat picture, there are, of course, caveats. Although the light curve of the supernova was fitted most cleanly to a shockwave-shell collision, there was a less clean fit to an explosion powered by 4.5 solar masses of radioactive nickel-56. However, this would represent an unusually large abundance of nickel for a core-collapse supernova. Yet, modeling by Ken'ichi Nomoto and colleagues does suggest that such large yields of radioactive nickel are not outside the realm of possibility, and the chance exists that a sizable fraction of the visible output of supernovae such as SN 2006tf is powered by radioactive decay.

SN 2006gy peaked at a greater luminosity than SN 2006tf, and this in turn peaked at a higher luminosity than SN 2008iy. Why is there this difference? Most likely this relates to the manner of the collision and the distance at which this collision began after the supernova occurred. Modeling suggests the collision that powered the more brilliant supernova SN 2006gy occurred at a smaller distance from the supernova than SN 2006tf and was driven by a faster moving blast wave. The shock wave in SN 20006tf was measured at 2,000 km/s compared to the 4,000 km/s shock of SN 2006gy. The greater kinetic energy in the SN 2006gy blast meant that there was more energy available to power the luminous output of the explosion – hence the brighter supernova.

Just to be different SN 2008iy was embedded in an optically thin (nearly transparent or translucent) clumpy field of hydrogen rather than a thick shell of matter. Consequently, the energy available from the supernova shockwave is more diluted in a succession of patchwork collisions. Thus the supernova brightens more gradually and to a lower level than supernovae that occur within a thick shell. The most illuminating feature of SN 2006tf is the interaction between the shockwave and a dense shell of material, and the subsequent impact with the clumpier, older wind of the presumably LBV star. This sort of event offers a clear insight into the behavior of the star prior to its cataclysmic end (Fig. 3.6).

In 2008 an even more luminous explosion occurred that bore the hallmarks of a collision between a dense shell of material and the shockwave from the supernova. Discovered by the ROTSE supernova verification project (RSVP), SN 2008am emitted an estimated total radiated energy of 2×10^{44} J and peaked at an absolute magnitude of −22.3, making it brighter than SN 2006gy and SN 2005ap. The supernova brightened over a modest period of 34 days. The lack of a P-Cygni profile or of intermediate width emission lines characteristic of fast-moving material in early times indicated that the visible energy was radiated from a photosphere that lay beyond the expanding shockwave of the explosion. At these times the shockwave was still traversing the inner part of the shell. In support of this, narrow emission lines in these early spectra are characteristic of hydrogen flash-ionized by the shock breakout from the exploding star.

114   Extreme Explosions

Fig. 3.6 SN 2008am, SN 2006gy and SN 2006tf. Both SN 2006gy and SN 2008am showed abundant evidence for collisions between their supernova shockwaves (*pink*) and surrounding (*circumstellar*) material. The shell (*graded pink-mauve*) surrounding SN 2008 (*top*) lay close to the star at the time of detonation, leading to a fast rise in light as the kinetic energy in the blast wave was rapidly converted to visible light. SN 2006gy (*lower image*) occurred inside a broader, but similarly massive, shell. The energy of the shockwave was more attenuated in this case by the more distant and extended nature of the shell, leading to a slower rise to maximum light and an overall lower absolute peak magnitude. SN 2006tf had a shell with a similar mass to SN 2006gy, with the shell lying at a similar distance to the supernova. However, the shockwave was slower moving than seen in SN 2006gy. Outside the shell lay a region of very clumpy gas

The presence of intermediate width emission lines of hydrogen in later spectra indicated that an outward moving shock was traveling at approximately 1,000 km/s. Again, like SN 2006gy and SN 2006tf, this is a relatively slow speed, indicating that the shockwave from the supernova was strongly attenuated through a collision with a dense circumstellar shell of matter. The total energy output was estimated at 10^{44} J from the light curve – a rather substantial figure.

Despite differences in the rise and decline from peak luminosity, SN 2008am showed many similarities with SN 2006gy. And the early time spectra of 2008am seem broadly equivalent to those of 2006gy. These early spectra seem to mark a time in the explosion when the shock wave was still embedded deep within the shell of matter that surrounded the star. In both supernovae, late spectra revealed a phase when the shock wave had moved forward enough to leave behind the hot, radiating surface of the supernova – its photosphere, where hydrogen ions recombined with free electrons.

Overall, the rise time for SN 2008am was suggestive of a smaller but much denser and more opaque shell. The resultant light curve reflected a more rapid conversion of kinetic to light energy as the shockwave collided with the dense CSM than was seen in the explosion of SN 2006gy (34 versus 70 days). The radii of the shells in SN 2006tf and SN 2006gy appear similar to one another, but the mass of the shell surrounding SN 2006gy was larger, hence the subsequently more powerful surge in luminosity when the shockwave arrived.

Thus the differences in the light curves of all three supernovae – and more distantly SN 2008iy – can be explained by the geometry of the matter surrounding each explosion. The shell surrounding SN 2006gy had a large mass but was present at a large radius. This produced a broad, highly luminous explosion. The smaller, less massive, but more opaque shell of SN 2008am powered a brighter but faster declining curve. The more distant shell of lower mass in 2006tf produced a less brilliant explosion. Unfortunately there were no early spectra of this explosion, so some aspects of the light curve were inferred. Finally, the clumpy but very extended CSM surrounding SN 2008iy produced a slowly rising curve. All four supernovae in turn illuminated many of the potential differences and similarities between collision-powered supernovae, revealing how the geometry of the circumstellar shells or clouds affects the light curve of these explosions.

The light curve of each of these supernovae suggested that a substantial amount of mass was shed by the progenitor in some form of giant pulse and that this pulse of matter was in turn embedded in a more classical LBV wind. What remained unconfirmed was the nature of the progenitor. A substantial dent had been left in the couch, suggesting a massive person had once sat there, but without the person a certain amount of inference must be drawn. LBVs have the mass, the prodigious and often violently impulsive rates of mass-loss and the overall temperament needed to shape their environment in the manner that was observed. You don't need to see the storm to know how the flood occurred.

Finally, if we consider the alternative mechanisms that may power these brilliant explosions we encounter problems. To drive the observed output of SN 2008am by radioactivity would require a preposterously large amount of nickel-56 to be synthesized in the explosion, 19 solar masses in total. The same is true for SN 2006gy: 22 solar masses. These quantities simply cannot be produced by even the largest feasible pair instability events in the modern universe; and even more importantly, the light curves of these explosions cannot be matched adequately by the decay of nickel-56 and its daughter nucleus, cobalt-56. The radiated energy must come from a collision, with radioactive decay being considerably more modest in its contribution. For the collision to have the desired effect the material must lie close to the star so that the blast wave intercepts it early and converts the bulk of its energy through the collision, rather than ameliorating it over a great distance.

Finally, a model involving a magnetar (Chap. 9) that blasts energy into the expanding shockwave was also considered by the authors of the 2008am paper. This has some success in matching the light curve of the explosion and cannot immediately be discounted. However, the presence of emission lines in the early spectrum of the explosion is not readily accounted for by the magnetar model, which involves delivering energy as the magnetar spins down. The magnetar model has some beneficial features but not enough to account for the light curve and spectrum of SN 2008am in its entirety.

These exceptionally luminous supernovae are rare, with an estimated rate of approximately 2.6×10^{-7} events per cubic megaparsec per year. Moreover, these supernovae are extremely heterogeneous in nature, with some being hydrogen rich, some hydrogen deficient, some showing evidence of collisions between shells of matter and some not. As such, more than one mechanism seems likely to explain their prodigious power.

Conclusions

As recently as a few years ago it was thought that the life of the most massive stars was simply an extension of the models we had for their less showy cousins. The scenario was that a very massive star became an LBV, then lost mass through powerful stellar winds to become first a Wolf-Rayet star and then death as a supernova. The only question was whether the remnant was a black hole or a neutron star. There existed the question of whether primordial stars could have such large, initial masses that they died through the pair instability mechanism. However, it was not generally thought that stars in the modern universe would retain sufficient mass to pursue this path. Observations now make it clear that the stellar evolution of the most massive stars is not as straightforward as was originally proposed, and that significant problems exist with the current models.

Looking at these issues again it is clear that stellar evolutionary models are far from complete for the most massive stars. SN 2005gl marked the death of an LBV with at least 50 solar masses. It had a high metallicity (or at least its host galaxy did), so why wasn't the progenitor degraded to a Wolf-Rayet before it died? SN 2006tf, SN 2008am and SN 2008iy all appear to be the products of LBV progenitors, but they have unknown masses at death. SN 2006tf, SN 2006gy and SN 2008am show some features expected for pulsational pair instability. Therefore, the plum pudding mix in which these supernovae swirl features considerable uncertainties and apparently inherent contradictions. If metallicity relates to mass loss as predicted, then why was the progenitor of SN 2005gl so large at death? If it began life very large, why did it die as an uncontested core-collapse supernova and not as a pulsational pair instability explosion as theory might suggest? And penultimately, if SN 2006gy and its siblings are pulsational pair instability events, then what sort of mass did they start out with to die this way? Finally, what sorts of stars give rise to Wolf-Rayet objects?

Some stellar clusters with stars of over 80 solar masses have WR stars embedded within them. Why are these stars of the WR flavor and not LBVs? Indeed, why are LBVs so rare? Metallicity and rotation are ruled out as possible (at least universal) solutions, as are the presence of binary companions in most situations. However, is the mystery of what makes a massive star an LBV and not a WR star merely a question of timing or a combination of different factors – some of which are mentioned above? Alternatively, we now know that 60 % of O-class stars are close (spectroscopic) binaries. Are LBVs the products of stellar merger in these systems (or in dense clusters), or are WR stars the outcome of rapid rotation and mass transfer between the stars in a binary star system? Many more observations and population studies are needed to clarify these issues.

Observations of the most massive stars raise many profound questions regarding stellar evolution. Have the stars we observe exploding while in a luminous blue variable stage simply evolved from much more massive progenitors than those we see transiting a Wolf-Rayet phase? Conversely, do less, but still very massive, stars simply evolve into WR stars before exploding, or can stars explode while LBVs only where stars have merged to create these monsters? The only solution must come from observations – however difficult this may be with such a limited sample. Can we show that the most massive LBVs become pulsational pair instability or pair-instability explosions, while the less massive ones explode as LBVs through a core-collapse route? Finally, will observations clearly show that the least massive LBVs transition to WR stars before undergoing core collapse?

This leaves metallicity as the joker in the pack. No one really agrees on the impact of metallicity on stellar life or death. Perhaps, this is a case of juggling one variable too many: metallicity; mass; mergers and rotation may all be factors that contribute to the fate of these massive stars. Perhaps we simply need to look at the effect of stellar mass and rotation on the evolution of the core before tackling the other variables. Get that out of the way before we then juggle the contradictory data – and the contradictory models – on the role of metallicity.

4. Collapsars, Hypernovae and Long Gamma Ray Bursts

Introduction

We begin our journey with chance observations by the Vela spy satellite in 1973. A well-trodden path begins with the discovery of short bursts of high energy radiation. Fortunately for us these were not the signature of nuclear detonations on Earth but instead marked the violent destruction of something other-worldly. What that something was remained a mystery for three decades.

Vela's detected bursts lasted from less than a second up to several hours, with the majority lasting on the order of a minute or so. Later observations divided the bursts into two camps: short and hard versus longer and softer. The duration of each class of burst was somewhat arbitrarily divided at 2 s. Until the launch of the BATSE satellite in 1991, this was about all that could be said for these bursts. They appeared to come from all directions with a frequency of roughly one per day, but whether their origin was local, galactic or extra-galactic remained a mystery.

From Star Wars to Star Death

With little information to go on, several possible explanations arose to account for these enigmatic eruptions. These ranged from an enigmatic stellar explosion through to antimatter detonations in some form distant empirial galactic war. Leaving the more lurid explanations to one side, during the 1980s the majority of astrophysicists assumed that gamma ray bursts originated within a 100 parsec (326 light years) range. They were most likely associated with old neutron stars, thus limiting the amount of energy that was needed to create them. However, data from the BATSE

120   Extreme Explosions

satellite demonstrated that these explosions had an anisotropic distribution; that is that they were evenly spread out in every direction.

This observation implied the bursts originated outside the disc of the galaxy, and with a paucity of faint bursts, we appeared to be seeing all the way to cosmological distances, thus all but ruling out models involving the galactic halo. Were these explosions to originate in a smaller sphere of space, they would form a flattened distribution associated with the galactic plane. However, there was no indication that they were located here, nor in the outlying galactic halo. At this point notions of fighting rebel forces or even the whimper of an asteroid fatally crashing onto a neutron star were ruled out. The energy required for these bursts was considerable, and something altogether more exotic was required to explain them.

With data continuing to stream in, and the confirmation that bursts arose at cosmological distances, the race was on to produce a model which successfully reproduced the observed ferocity of these events.

Beppo-SAX to the Rescue

The final pieces of the puzzle arrived courtesy of the European X-ray satellite Beppo-SAX. Launched in 1996, this prodigal child broke the seal on gamma ray bursts (GRBs), revealing their true identity. Prior to its launch, activity in the field of gamma ray bursts was clearly tightly constrained. Little information could be extracted from a burst of gamma rays lasting seconds, or at best a few minutes. The prolonged notification time and the consequently sluggish interval required to slew to any available ground-based telescopes, or even Hubble, to observe the bursts was neigh on impossible. Even if the technology existed to report and activate secondary telescopes in the limited time available, the low resolution of the active gamma ray observatories restricted what, if anything, could be extracted from the location of the fading burst. Thus the proverbial snail easily outpaced the developments in the field. The arrival of Beppo-SAX changed all of this.

The key to the success of Beppo-SAX lay in its combination of a low-resolution gamma ray detector and a higher resolution X-ray 'scope. Using these two detectors in combination allowed a precise location for the gamma ray emission to be established. The first success was GRB 970228. Eight hours after detection with the onboard gamma ray detector, the satellite was rotated so that any X-rays produced in the burst could be collected and the location of the explosion resolved. The X-ray detector had a resolution of less than 5 arc min, or a 70th of a degree. Using the narrow window yielded by Beppo-SAX's X-ray detector the William Herschel telescope located the fading afterglow of the explosion at visible wavelengths within 4 days of the initial discovery.

Registering at a dim magnitude 21 on March 1st, this paltry whimper of light made its origin clear through its steady decline over subsequent days. By March 8th the glimmer of light had faded to magnitude 23. Three weeks later, on March 26th, only the Hubble Space Telescope was able to pick up the marginal glow at a dismal magnitude 26. As the candlelight faded, the William Herschel and later HST detected a faint background glow, suggestive of a host galaxy lying at great distance from Earth. With the glow insufficiently strong, researchers held back from claiming a definitive capture of a host, but clear distortions in the fading light were highly suggestive. However, it would take a burst 3 months later that year to clinch the deal for a galactic origin for long bursts.

A tranche of papers in the journal *Nature* heralded the new era in gamma ray burst research. In February Enrico Costa and colleagues reported on the fading X-ray afterglow of GRB 970228, while Johannes van Paradijs reported the fading optical emission and the discovery of a probable host galaxy for this burst. Finally, in May 1997, Titus Galama discussed the structure of the fading optical afterglow and the impact these observations have on determining the nature of the explosions themselves. Galama's observations ruled out some of the simpler models for gamma ray bursts, putting some constraints on the types of physical models that could apply to them.

Like an artist developing a painting with every stroke, observations of GRB 970508 added further brushstrokes to the emerging picture. The first measurements of the radio emission from a gamma ray burst were made, indicating that the bursts had a complex physiology with rapid variations and unexpected differences in the rate of decline of the burst with wavelength. Observations at radio wavelengths by the VLBI (Very Long Baseline Interferometry) also allowed the determination of parallax, which confirmed the compact nature of the fireball that generated the GRB. Radio observations also confirmed some motion of the source across the sky, as would be expected for a distant source.

Complementary radio observations of the May GRB by Dale Frail indicated that the fireball expanded rapidly over the first few weeks of observation, meaning that very high energies were likely involved in generating the afterglow. Observations of GRB 970508 were made at. 4.86 and 8.46 GHz, and although the alignment of the radio source detected by Dale Frail and co-workers with the X-ray afterglow was a little subjective, the odd variability in the source (up to 50 % of the total radio output) and its unusual spectrum argued for an association with the GRB.

One of the unusual features was the wavelength dependency of the rise time of the radio emission. The time to plateau was 3 weeks at 4.86 GHz but only about 10 days at 8.46 GHz. This peculiar trend in the rise time varying inversely with frequency continued at the shorter wavelength of 1.43 GHz. The implication was a source that brightened fastest at shorter wavelengths. This might be expected from an expanding and cooling ball of gas – a fireball. At first the ball is smaller, hotter and expanding faster, causing shorter wavelength emission. As the fireball cools, increasing amounts of the matter contained are able to radiate at longer wavelengths. However, since the fireball was now larger the time to reach a peak would take longer.

Perhaps one of the most crucial early observations was the consistent picture of X-ray afterglow emission, amounting to 40 % of the total gamma ray output when measured in the range of 40–700 keV (thousands of electron volts). The electron volt serves as a useful measure of particle energy and is preferentially used to describe and relate the energy in radiation with its source in the atom.

Although extended over a longer interval, it became evident that emission at X-ray energies contributed a sizable fraction of the total output of the burst. Likewise, optical emission amounted to a consistent 3 % of the gamma ray emission. This suggested a broadly reproducible mechanism operating within these bursts that generated their output. The observations by Galama also showed an interesting temporal variation in the pattern of decline of GRB 970228. Up to 6 days after the burst, the optical emission declined by a steep factor of 40. But subsequently, the decay rate slowed considerably. This could have been a consequence of the impact of an underlying galaxy – or perhaps it was the first hint that at least some of these bursts were associated with underlying supernovae.

GRB 970508 was significant in one other way. For the first time, optical emission was observed rising to its peak and then declining over a 2-day period following the GRB. The optical afterglow was detected at a very early stage – just 6 h after the burst was detected. It took the brilliance of the design of Beppo-SAX to make this possible. Such prompt observations allowed astronomers to constrain many of the models used to explain the bursts. The students were observing the artist at work. The patterns of changing light and shadow inherent in GRB 970508 were the brushstrokes of the expanding fireball associated with the GRB.

Certainly, there must have been considerable serendipity in unveiling this part of the adventure. Capturing the rise of the explosion to maximum light must have been a magical experience for those involved: this was a photographer catching a Mayflie mating on the wing in a series of images taken once per month over its entire life. However, forewarned is forearmed. With GRB 970228 in the bag, astronomers had a good idea of what to expect when the second opportunity came along that May. A loose analogy might be crossing the Alps in an overnight train. The first traveler awakes from a slumber while the train is descending its winding track into Italy. Overwhelmed with the majesty of his view, he telephones his successors to look out for more when they make the same journey. Those following then stay awake throughout, seeing the Jura Mountains, the Rhone valley, the high Alpine passes and finally the soaring Dolomites of northern Italy.

Of Fireballs and Jets

The pattern of decline in X-ray emission fitted a broad power law. These were in accordance with some fireball models proposed by Stan Woosley in his seminal 1993 paper describing the so-called collapsar model. However, these early observations severely constrained the scope of the models. The simplest fireball models were ruled out because they predicted a fast rise in visible emission. Instead, observations made between the 9th and 10th of May painted a picture of a more modestly expanding fireball.

As the radio observations constrained the nature of the host galaxy, Bond and co-workers reported the unambiguous discovery of a blue starburst galaxy underlying the May 1997 GRB, using the 200-in. Palomar telescope in California. Now, astrophysicists had both a handle on the explosions and evidence that at least one host burgeoned with active star formation. The collapsar model of Woosley received another nod. The massive stars generated by starbursts were becoming the likeliest hosts for these explosions.

Interestingly, despite the strong hint that these explosions might be jetted in nature, many of the early papers assumed the explosions created a spherical blast – a fireball. Were this the case, the evidence of very high, relativistic speeds would imply a total energy 100 times that of a conventional supernova – 10^{53} erg or 10^{46} J. Gregory Taylor (NRAO) even suggested that the total energy in GRB 970508 was as great as 10^{54} erg, or 1,000 times that of a core-collapse supernova. Such models were hard to explain, as the energies required were so high. What sort of central engine could generate 1,000 times the output of a supernova in a matter of seconds? To put this into perspective, 10^{54} erg or 10^{47} J is approximately the same as converting a Sun's worth of mass into energy. The universe is a violent place, but could it possibly be this violent?

However, there was a simple solution to the conundrum posed by the energy output. If the energy was beamed in a narrow direction – perhaps as little as a few degrees, then the energy would be reduced by simple geometrical considerations. And so, instead spherical blast waves gave way to the idea of tightly focused beams, and some of the more perplexing features of the X-ray afterglows could be more readily explained. One of the more peculiar elements in the afterglow was the so-called break in the light curves. This is illustrated with the X-ray spectrum shown in Fig. 4.1.

The jet-break feature is best explained using the geometry of a jet beamed towards the observer. The early, highly luminous

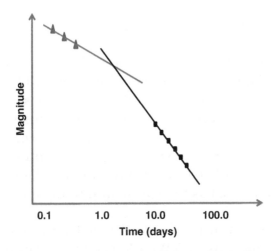

FIG. 4.1 The jet break. Early observations showed a shallow rate of decline in the X-ray luminosity, but after a hiatus, the luminosity then fell, with a much steeper gradient. This is hard to explain if the GRB has a spherical geometry. However, it is readily explained if the energy is beamed in the direction of the observer

portion is energy emitted from the head of the jet. As the jet advances, the emission from the head fades and lower energy emission from energetic electrons in the tail of the jet become visible. This is detected as a fast declining emission.

The nature of the afterglow was also contentious. In models involving a spherical blast, the afterglow was generated by synchrotron emission as the blast wave plowed into ambient gas surrounding the star. Synchrotron emission, you may recall, is a special kind of radiation generated by electrons spiraling in a tight magnetic field, such as is found in the blast wave of a supernova (Chap. 2). Indeed, the X-ray and optical afterglow of GRB 970228, especially its decay, appeared to fit a simple formulation of this model. However, contrary to the notion of a simple fireball, the optical and radio afterglow of GRB 970508 showed a slow rise to a plateau value before a decay set in. These differences could be explained by variation in the density of the gas surrounding the star, or by variations in the energy of the electrons present in the gas lying behind the shockwave. However, at the time a precise description of the environment that generated such emission eluded astronomers.

A Best Fit: The Collapsar Model

Let's look at the GRB scenario again and begin to place it in the context of stellar evolution. The GRB is generated by the collapse of a particularly massive star – the precise mass range is still unclear, but later observations will constrain this with a breathtaking precision. Somehow, the collapsing core generates two narrow jets that burst out of the star and generate gamma ray emission accompanied by a tail of lower energy emission. Most of the X-rays appear to be produced as the jets slam into material surrounding the star.

In 1993 Stan Woosley proposed a model involving a compact, massive star – most likely a Wolf-Rayet. Other, larger model stars had various problems that we shall look at later, but the principle objection to these was the likely loss of energy in the jet or fireball as it traversed the breadth of such larger stars. By contrast, a compact WR star has sufficient mass to produce a black hole, while being limited in its dimensions, thus allowing a jet to pass without much energy loss. In the model the WR star evolves a particularly large iron core, but the crucial feature is that the core – and the star as a whole – is that it is rotating quickly. Such fast rotation may be a consequence of a stellar merger or simply because a close binary partner, which is similarly massive rotates rapidly around their common center of gravity.

Once the core becomes unstable to gravity it implodes to become a black hole with a few times the mass of the Sun. Material lying along the rotation axes of the star is no longer supported against gravity and aggressively descends to its doom in the nascent black hole within a matter of seconds. This clears two opposing channels out of the stellar core. The remaining material forms a thick disc around the widening black stellar equator and begins a brief but sufficiently protracted phase of accretion that feeds material to the hole at a rate of a few tenths of a solar mass per second. Meanwhile, all this activity produces a boiling froth of neutrinos and antineutrinos around the polar axes of the black hole.

In Woosley's model, gamma rays emerge from this billion degree froth and then escape along the cleared rotation access of the dying star, illuminating a cone of space projected north and south from the star. This collapsar model, in different and evolving forms, has subsequently served as the cornerstone of GRB physiology – or, more precisely, the cornerstone of our understanding of *long* GRBs, those with outputs lasting more than 2 s that release softer (less energetic) emission. The short, hard GRBs appear to form by a different route.

In 1999 Woosley, working with Andrew MacFadyean, produced more realistic computer simulations of these explosions, which took into account rotation and variation in stellar mass. The most successful model involved a 15 solar mass helium star that evolved to core collapse. Such a star is a close match for many WR stars that have shed their hydrogen shells. Models involving lower metallicity were generally but not exclusively better at forming GRBs, as the mass of the silicon shell was greater. This drove more material onto the nascent black hole, powering a meatier pair of jets. Very low metallicity stars encountered problems in MacFaddyean and Woosley's models, as they retained much of their hydrogen-rich outer layer. This absorbed a lot of the energy of the jets and thus prevented their energetic escape.

These more massive stars appeared to generate softer radiation bursts; this was also true if the GRB was observed off-axis. Remember, the jets giving rise to GRBs appear to be very narrow (covering a few degrees of sky at most). Woosley suggested that a stellar collapse observed off-axis would display softer X-ray emission. This was something suggested several years later to account for X-ray flashes, or XRFs (see later). Similarly, some XRFs show more leisurely jets moving at lower fractions of light speed and again are manifest as XRFs. Once more these may be the consequences of jets erupting in larger stars.

Woosley's models incorporated three crucial features that were testable: compact massive stars occasionally gave rise to gamma ray bursts; the formation of a black hole was heralded by these explosions; and the emission was generated close to the black hole in some form of fireball. Other features of the model have been eroded with later and more precise observations, but the basic tenants have stood the test of time.

Woosley identifies three kinds of collapsar. A Type I collapsar is perhaps the best described in the literature. A Wolf-Rayet star began as a larger 40–95 solar mass star, but then evolved through to core collapse. By death the star has shed its hydrogen-rich envelope,

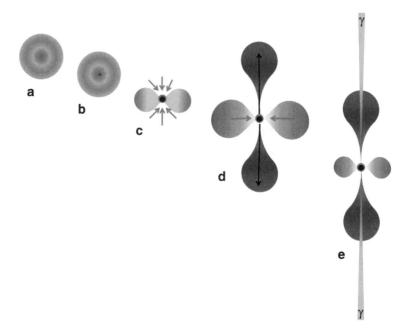

FIG. 4.2 The collapsar model. The generation of a long gamma ray burst. In (a), a massive Wolf-Rayet star becomes unstable when its core fills with iron. By (b), the core has collapsed to form a stellar mass *black hole*. At (c), the *black hole* rapidly accretes material along the polar axes, which is not supported by the rotation of the star. This clears two opposing channels. In (d), an accretion disc is feeding material into the *black hole* along the stellar equator, while two broad outflows of material are pushing upward. These may be driven by two narrow (1°–5° wide jets), which in (e) are generating internal shocks, which in turn produce the observed gamma rays (γ). The broader outflows may generate an observed supernova and/or a lower energy X-ray burst

and its mass was stripped down to a naked helium core weighing in at 15–40 times that of the Sun. Festering within the helium star the iron core became unstable, and a black hole formed through prompt collapse. This generated the conditions of the gamma ray burst that we have encountered already (Fig. 4.2).

A Type II collapsar may occur in less massive stars where the helium core was less substantial. The core collapses to form a neutron star that is initially stable against any further collapse. But the denser structure of the helium star drains the life out of the shockwave as it struggles outwards. An insufficient amount of matter is blown away into interstellar space in the accompanying supernova, with most of the rest tied down in the gravitational pull of the young neutron star. Unable to escape, the majority of the radioactive nickel minted in the explosion falls back onto the neutron star, along with a substantial portion of the oxygen and carbon shells of the former core. Weighed down by this additional mass, the neutron star implodes and a black hole forms. Continued accretion through a disc feeds the black hole and generates the jets that drive the gamma ray burst.

The time between neutron star formation and implosion appear to be on the order of seconds to minutes, depending on the rate of accretion. Any accompanying supernova is expected to be particularly weak, as the limited mass of material that does escape is likely to be deficient in radioactive nickel-56 – most of which is now plunging onto the neutron star.

Ken'ichi Nomoto has found that in some instances, such fallback events may shine sufficiently powerfully to be seen across intergalactic space, if the jet – and perhaps other processes – help mix radioactive nickel outward into the overlying shells, and away from the developing black hole. However, such mixing may be a rather capricious process, with a very variable outcome. Hence the results of fallback may be very different for two rather similar stars if one or more of their properties (such as spin) diverged over their lifetime.

Finally Type III collapsars are thought to be confined to the early universe. These early, Population III stars may have had particularly large masses – up to 1,000 solar masses. Such stars retained much of their mass up to the point of core collapse. When these stars finally blew, the helium core would have retained more than 130 times the mass of the Sun. Unlike Type I and II collapsars implosion occurred at an earlier stage in the stars evolution. When oxygen ignited the core became unstable and collapsed. As these stars imploded the majority of the star's material would have rained downwards to produce a particularly large black hole with perhaps 100 times the mass of the Sun. Evidence for these monsters is limited to theoretical papers. At the moment it isn't clear how many of these occurred in the early universe, but examination of the chemical composition of Population II stars implies that their role may have been fairly limited.

Hypernovae and Hyperbolae

In 1998 one explosion occurred that revolutionized our understanding of the stellar calamity that gives rise to a GRB – although if you read the cautiously worded papers of the day, you could have been forgiven for underestimating the magnitude of the discovery. In a nearby spiral galaxy SN 1998bw appeared. It bore many unusual features for membership of the Type Ic fraternity. Although clearly belonging to the class of supernovae deficient in hydrogen and helium, this explosion showed very broad emission features and unusually strong radio emission, both characteristics of material moving at a sizable fraction of the speed of light. Essentially simultaneously, GRB 980425 irradiated Beppo-SAX, triggering the hunt for an afterglow. Measurements of the emission from the GRB suggested that at most 10^{-5} solar masses of material, or a hundredth the mass of Jupiter, was ejected in the jets – a rather unspectacular value; plus, the peak gamma ray emission was four orders of magnitude less than many of the previously observed "cosmological" bursts. Overall, this was a fairly wimpy gamma ray burst.

Meanwhile, SN 1998bw displayed a number of peculiarities that led some to quickly suggest that this explosion might be more than just spatially coincident with the GRB. They might be causally related. In terms of its peak luminosity SN 1998bw would be regarded as rather underwhelming by today's standards, clocking in with a peak absolute magnitude of –19.5, but at the time this was many times more luminous than the majority of observed Type Ic events. However, given that this modestly high luminosity was driven solely by matter produced and ejected in the explosion itself, the implication was that the explosion was unusually energetic. Two pieces of evidence were important here. The first was the high ejecta velocities, while the second was the above-average peak luminosity. Together these implied that 0.5–0.7 solar masses of radioactive nickel had been synthesized – ten times that seen in other core-collapse supernovae – and more in line with thermonuclear Type Ia events. Finally, the strong radio emission and broad emission lines in the spectra implied particularly high velocities in at least a sizable fraction of the ejected material.

SN 1998bw *was* unusually energetic, but why?

On their own these features in the supernova would have sent minds scurrying in search of answers. However, it was the nearly coincident location of the supernovae to a gamma ray burst seen days earlier that caught the most attention. SN 1998bw lay within the error box of GRB 980425, suggesting that the two events were related. A year earlier SN 1997cy also shared a spatial location with another GRB. However, this supernova was poorly constrained temporally by observations, nor was the spatial coincidence of the gamma ray burst particularly good. (Indeed there were two possible bursts associated with this supernova at different locations in the sky.) Without clear observations, SN 1997cy lost out to SN 1998bw as the first supernova candidate to be linked to a long gamma ray burst.

Although, there certainly wasn't a cast-iron case linking the two events, the peculiar kinematics of the supernova would readily be explained if the GRB had occurred in the star which then somehow triggered the supernova. Various papers documenting the GRB and supernovae nervously linked the two events, but perhaps with understandable reticence they avoided claiming one directly led to the other. However, careful scrutiny of the explosion revealed that the architecture of the supernova matched predictions of Stan Woosley's collapsar model. Although not directly observed, the progenitor was clearly a compact WR star, as the spectrum displayed neither hydrogen nor helium.

So, had the elusive supernova-GRB connection been bagged, confirming Woosley's conjecture? At the time a somewhat oxymoronic definite "perhaps" would have been the response. Nowadays, most researchers would stake their reputations on it. However, it would fall to later GRB discoveries to make astrophysicists confident to claim the link was confirmed.

At the time SN 1998bw was documented the term *hypernova* was coined by Koichi Iwamoto to describe the apparent additional oomph in the explosion. Initial estimates placed the energy in the expanding shock wave at 10^{52} erg or 10^{45} J – an order of magnitude above a typical core-collapse event, and far in excess of most of the Type Ic supernovae that had previously been observed. Although such feats might just be feasible in the environment of a supermassive black hole, there was a significant question as to whether an individual star could create so much energy in such a short interval of time. Thus, even at this stage the idea of a "hypernova" was questioned. Even at this point the unprecedented and difficult to explain energies could best be accounted for by beaming of energy across our line of sight. Thus 10^{54} ergs suddenly drops to more manageable $10^{51\text{-}52}$ ergs, placing these explosions in line with typical supernovae. However, what remains spectacular, even in this more bashful scenario, are the velocities of a proportion of the debris. Launching matter at up to 99 % of light speed, the term "hypernova" stuck in many astronomers' minds and soundly resonated within the mass media.

However, names are like ghosts and have a nasty habit of coming back to haunt you. As more data has streamed in and the GRB-supernovae have lost their top spot for shear unbridled energy, the term "hypernova" has seemed, well, overstated. Moreover, the term "hypernova" has been variously applied to all sorts of unusually bright supernovae, from hypothetical Population III blasts to gamma ray bursts, pair-instability explosions, and, well any particularly bright supernova. Thus although superficially appealing, the term "hypernova" is so vaguely and widely applied as to make it effectively meaningless, so it shall appear no further.

So, why were authors so reticent about making a formal link between GRB 980425 and SN 1998bw? In part this is the scientific process –evidence first and claim second. Secondly, with most X-ray data coming in with low resolution, there was sufficient breadth in the error box of the GRB to make an unambiguous link with the supernova. Thirdly, the GRB was unusual in that it appeared to be very weak, so where a link was proposed between the two explosions, such as in the paper of Ken'ichi Nomoto's group, special circumstances were invoked to account for the possible association for a GRB with a supernova.

GRB 980425 was unusual and perhaps was not functionally equivalent to the explosions seen at cosmological distances. Giving a flavor of this reticence, in Iwamoto's *Nature* article he says, "If this supernova and GRB980425 are indeed associated, the energy radiated in gamma rays is at least four orders of magnitude less than in other gamma ray bursts, although its appearance was otherwise unremarkable: this indicates that very different mechanisms can give rise to gamma ray bursts." Looking back at this, it seems a peculiar stance given our current knowledge of these bursts, yet the scientific view is to describe then to explain. The claim to have found a possible supernova-GRB connection rested on a single event. The concept of reproducibility came to the fore – as did Carl Sagan's adage "extraordinary claims require extraordinary evidence." Thus it was perhaps understandable that the final clincher linking GRBs to rare core-collapse supernovae would come a few years later.

Yet despite the reluctance to formally link some supernovae to GRBs, Iwamoto and co-workers produced a rather flawless model of the supernova, involving a progenitor star composed of carbon and oxygen (CO) – and just the sort of star Woosley had proposed 5 years earlier as the source of GRBs. At its death the model star weighed in at 13.8 times that of the Sun and was derived from a more massive progenitor of 40 solar masses that had undergone extensive mass loss. In line with Woosley's model, this stripped stellar core was probably rotating very rapidly, and it was this high angular momentum that helped the corpse generate a final blaze of activity – the GRB.

Let's pursue the model. Throughout its life stellar winds had ripped and stripped away at this massive star until it was reduced to a bare CO core – a Wolf-Rayet star. At its ruin, the core of this stripped object collapsed, generating a black hole of 2.9 solar masses. Jets erupted from the maelstrom, swirling around the gaping maw of this rapidly rotating black cauldron. These jets then tore through the star, generating the observed GRB. Flanking the jets, the expanding blast wave ripped viscera from the stellar interior that was rich in radioactive nickel-56. This generated a luminous supernova whose output peaked close to that of SN 1998bw. Furthermore, the record-breaking photospheric velocity of 40,000 km/sec was reproduced by the core-collapse model following these parameters.

It must be emphasized that 40,000 km/sec is more than 10 % of the speed of light, and these kinds of speeds were utterly unprecedented in supernovae observed at the time. Even at late stages the spectra of SN 1998bw revealed outward velocities exceeding 10,000 km/sec. At the time of publication these high velocities could not be explained with standard core-collapse scenarios. An altogether more energetic root underlay them. This was the formation of a stellar mass black hole, accompanied by jets of matter moving at a sizable fraction of light speed.

Iwamoto and colleagues proposed that these relativistic jets erupted from the stellar surface, generating both these high photospheric velocities and the subsequent gamma rays. Clearly diverging somewhat from the model of Woosley, they proposed that the gamma ray emission was generated as the jets slammed into the surrounding circumstellar medium, resulting in a powerful synchrotron emission. Woosley had suggested that the gamma ray emission was generated close to the black hole in a hot bubble of matter and antimatter. Iwamoto's model had gamma rays produced much further afield. Further evidence would subsequently accumulate, casting doubt on both these explanations, but that aside, the basic collapsar model seemed to work.

Like Woosley some years before, Iwamoto suggested that the progenitor of SN 1998bw and GRB 980425 achieved this remarkable death as a result of high angular momentum or strong magnetic fields. The presumption was that this was the result of the merger of the progenitor with a companion star. This is something we have already seen. Therefore, the implication was there in the data: SN 1998bw was associated with GRB 980425. In an interesting twist to the collapsar model, Jonathon Granot, working with Stan Woosley, proposed that the properties of GRB 980425 and the accompanying supernova could best be explained if the GRB was observed just off the axis of the jet. Thus the somewhat underwhelming explosion energies of GRB 980425 were a consequence of beaming and the associated relativistic effects, rather than the GRB being intrinsically faint. GRB 980425 might not have been so wimpy after all.

As the millennium turned, more GRBs were captured that displayed the distinct signature of an underlying supernova – the presence of high levels of metals synthesized in the explosion and then ejected from the collapsed core in the burst. However, the picture was not uniformly clear, nor was the chronology of the observed events. Did the supernova precede the GRB, or was it the other way around? Moreover, another couple of questions arose. In what proportion of supernovae might we observe a GRB, and conversely, in what proportion of GRBs do we expect to observe supernovae? In both instances, is it all or some? If only some, why do some GRBs give rise to, or are accompanied by, supernovae, while others do not?

As the wheel of time turned and the new millennium began, improvements in computational power allowed the collapsar model to be refined and the processes of stellar death to be re-examined in considerably more detail. In particular, with a reasonable nod to computational time, the processes involved in taking a star up to its point of death and then modeling what happens when a black hole is formed could be simulated in two dimensions on supercomputers. In the late 1990s and early part of the last decade, Andrew MacFaddyen, working with Stan Woosley and others, produced the first two-dimensional models of the events unfolding before and during a GRB. The models incorporated contemporaneous observations that some GRBs appeared to originate from Wolf-Rayet stars.

The refined collapsar model incorporated an imploding core that generated two narrow jets which punched outwards from the star. Initially, the jets were rather broad, measuring 10–20° across. These broad jets generated shockwaves that tore around the star, blasting material outwards at the stellar equator and generating a supernova. Meanwhile, these broad jets cleared a path around the rotation axes of the star, which allowed much narrower and faster jets to penetrate the tattered stellar remains. These energetic jets were only a few degrees across, and crucially, unimpinged by the matter from the star, could accelerate to speeds close to that of light.

The revised models disposed of earlier suggestions that the gamma rays were generated near the black hole. They also cast doubt on models involving the collision of jets with circumstellar material. Instead the jets made the gamma rays directly. The mechanism is fairly complex but essentially boils down to the transfer of energy from particles of matter to photons of light through close encounters and collisions. In essence, as the jets move out from the surface of the defunct star, collisions occur in the jet between packets of matter moving at slightly different speeds. These internal shocks subject ambient photons of light from the star to further collisions. During these violent encounters photons pick up energy from fast-moving electrons and protons. These particles lose energy, while the photons gain it, raising their energy levels to those of gamma rays. The process is known as inverse compton scattering.

Around the highly energetic core of the jet, slower moving material can drive photon energies to lower, but still extreme values, generating an associated burst of X-rays. Finally, minutes to hours later, the jet collides with material surrounding the star. This can be interstellar gas, or gas and dust ejected from the star in earlier epochs of mass loss. Either way, the collision of the jet with this interstellar debris converts much of the jet's energy into X-rays and the accompanying optical and radio emission, observed by Beppo-SAX, Hubble, William Herschel and the VLA.

Over the ensuing years, this model has held up very well to scrutiny. Further observations have added evidence for the presence of magnetic fields in the jets, and in at least one instance a contemporaneous optical glow was found associated with the GRB. Like any good model, these observations have added to it rather than discounted it and have served to illuminate the inner workings of the fiery birth pangs of black holes.

Reconstructing the Supernova-GRB Connection

Let's return to the supernova-GRB connection and repeat the question, are all *long* GRBs associated with supernovae? Well, the evidence appears to say, "No." By 2010 only a handful of GRBs were unequivocally known to accompany core-collapse supernovae. These were: SN 1998bw and GRB 980425 at a red-shift of 0.0085; SN 2003dh and GRB 030329 at a red-shift of 0.17; SN 2003lw and GRB 031202 at a red-shift of 0.1; SN 2006aj and GRB 060218 at a red-shift of 0.033; SN 2008hw and GRB 081007 at a red-shift of 0.53; with another three (e.g., GRB 011211) showing spectroscopic evidence for supernovae associated with GRB afterglows but without the unambiguous detection of the supernova itself. Indeed, a month prior to GRB 980425, GRB 980326 displayed some interesting features in its spectra – red bumps that could have been interpreted as the light from an underlying supernova. However, without observations of a host galaxy, or even a convincing redshift, the detection of this curious afterglow went without much further scrutiny.

However, the supernova-GRB connection was to re-emerge, butterfly-like, in 2001. GRB 011211 was once again discovered by Beppo-SAX, and lasting more than 3 min in length, it was a record for the Italian-Dutch satellite. GRB 011211 delivered a punch more than twice as strong as GRB 980425, with an estimated total energy of 5×10^{45} J and an outflow velocity tracked at 25,800 km/sec. Reported by James Reeves and colleagues (University of Leicester, UK), the explosion lay at cosmological distances (a red-shift of 2.14), and the X-ray spectrum suggested the presence of magnesium, silicon, sulfur, calcium, argon and possibly nickel (although evidence for the latter was rather more ambivalent). Such elements are typically the chemical fingerprints of core-collapse supernovae.

This broth of elements was seen moving away from the burst center at great speed. Spectra indicated that the abundances of these elements reached ten times that seen in the Sun. This indicated that their origin lay in a supernova accompanying the GRB. Moreover, Reeves suggested that fluctuations in observations at the red-end of the spectrum at 12 h after the burst were due to physical inhomogenities in the circumstellar matter 40–125 AU distant from the moribund star. Later spectra, taken after the typical "jet-break" (Fig. 4.1) indicated a smoother profile, indicative of a more homogeneous circumstellar environment. Reeves went on to suggest that these differences reflected variations in the history of mass loss from the star in the run-up to its implosion and that these shared a close synergy with observations made of other massive stars.

However, soon after the announcement and publication of the results, numerous other researchers, including Stephen Holland, became critical of the methods used to analyze the spectra, and it was suggested that the XMM-Newton data was based on artifacts. Thus the claim that this GRB was associated with an underlying core-collapse supernova was filed as unproven.

Yet the game was not over, and 2001 delivered a second possible GRB fostering a link to an underlying supernova. Nursing a confusingly similar date of birth, GRB 011121 (note the subtle difference) was clearly associated with SN 2001ke and occurred at a lower red-shift of 0.362 than GRB 011211. The supernova emerged in the declining optical emission of the GRB 12 days after the GRB had peaked – leading principal author Peter Garnavich (University of Notre Dame) to claim the prize for capturing the first cosmological supernova-GRB connection. In parallel Shri Kulkarni and fellow workers from Caltech made a parallel claim. Was this the first supernova unambiguously associated with a gamma ray burst?

Photometric analysis, carried out by Garnavich's Dutch team, indicated that the supernova was unusually blue, leading Garnavich to suggest this blue continuum resulted from the interaction of the supernova blast wave with a dense circumstellar shell (see last chapter). "This is the best evidence to date that classical, long GRBs are generated by core-collapse supernovae," said Garnavich at the time.

Interestingly, as Garnavich noted, many of the claims for supernovae rested on the discovery of a red bump in the declining light curve of the GRB. Garnavich's supernova was distinctly blue, leading him to infer that supernovae associated with GRBs were a rather diverse bunch of objects. Retrospectively, Stan Woosley made the interesting point that the "blueness" of the explosion might not be due to the physical interaction of the GRB jet with its ambient surroundings; rather this was an effect of seeing more deeply into the core of the star, down a corridor cleared by the jet.

In 2003 the discovery of SN 2003jd added a further final twist to the supernova-GRB connection. The supernova was a particularly energetic Type Ic event, with clear evidence of material ejected away from the star in two opposing high-velocity jets. Examination of spectra revealed a curious double-peaked feature in the oxygen and magnesium emission lines. The double peak suggested two opposing features moving in opposite directions. A similar structure was noted in the spectrum of SN 1996aq – a rather run-of-the-mill Type Ic explosion, suggesting that such strong asphericity may be typical of Type Ic events in general. However, the deduced velocities for SN 2003jd were notably higher.

What, then, can we say about those supernovae that are associated with long gamma ray bursts? Most appear to be associated with Type Ic supernovae, which are themselves strongly aspherical in shape. Compared to the majority of other Type Ic events, these are more luminous than average, but less so than the brightest explosions thus far observed (and described in Chaps. 3 and 8).

The luminosity and high outward velocity of the debris implies that 0.5–1.0 solar masses of radioactive nickel-56 is synthesized in the explosions. These explosions are very rare, occurring in less than 1 % of all core-collapse supernovae. The progenitors are most likely fast-rotating Wolf-Rayet stars with initial masses in the range of 30–40 times that of the Sun. They may form half of a tight binary pairing of stars, something that is common in their O-class progenitors, hence the apparently high rotation.

Other long GRBs seem not to have been associated with underlying supernovae. A number of very sensitive searches were done with Keck and Hubble of the locations of some of these. In many of these there was a "non-detection" of an underlying supernova to luminosities well below SN 1998bw. One rather interesting example is GRB 020124. In another event captured by Beppo-SAX, the optical emission from this burst faded with extreme rapidity and was only narrowly caught. No underlying supernova was identified, and was it not for the prompt observations of Edo Berger (Caltech) this long GRB might have been classified as dark – a GRB born without any accompanying optical emission or afterglow. Some long GRBs are intrinsically faint, and the underlying reason or reasons for this are unclear.

Although the absence, or more precisely the non-detection, of supernovae is not definitive, it is at least suggestive that any underlying supernovae must be unusually faint. However, such feeble supernovae are not without explanation in the models of Woosley and others and may be explained by the process of fallback (Chap. 2). Therefore, the jury remains out as to whether these explosions are supernovae-free or simply supernovae-faint. What is clear is that GRBs are very rare phenomena in comparison to core-collapse supernovae. Indeed, they may be as rare as one per million core-collapse events; therefore, we may then ask what makes a GRB, and what's lacking in lesser beasts?

Supernova: Or Supranova?

Could some GRBs be of Woosley's Type II variety? If so, just how much delay is possible between the formation of the neutron star and the black hole? Let's paint a picture of a fast-rotating

Wolf-Rayet star, its core increasingly unhinged, struggling for its survival, clinging on for dear life. Suddenly, electron degeneracy pressure, the repulsive effect of electrons within its iron core, fails, and the core implodes. A neutron star forms, and the subsequent shockwave plows outward, ripping the star asunder. A familiar plot appears to be playing out on the cosmic stage.

The shockwave travels outward, dragging filaments of metal-enriched gas to distances exceeding the orbit of Neptune, and suddenly, wham! The neutron star implodes, forming a black hole. Two opposing jets of metal-rich gas blast outward at close to the speed of light. A few days later they catch up with the outward moving shell, generating a furious burst of X-rays that bear the chemical signature of the earlier supernova. This is the *supranova* – a core collapse that stutters before its final calamity. A stellar death separated in time from the ultimate fate of the collapsed core.

In the case of the problematic supernova associated with GRB 011211, it was suggested that the supernova preceded the GRB by 4 days. This seemed unlikely. However, in August 2002 GRB 020813 was detected. Again, and this time with convincing data from the Chandra X-ray Telescope to accompany it, evidence was found suggesting that the jets from the gamma ray burst had caught up with and illuminated a shell of metal-enriched matter ejected some 60 days prior to the gamma ray burst. The shell appeared to be expanding at one tenth the speed of light, or 30,000 km/sec – a speed comparable to, but slightly faster than, the out-flowing matter from SN 1998bw. However, in the case of SN 1998bw, there was a clear synchrony between the supernova and GRB 980425; and the velocities of the expanding matter tie in with the jets launched that generated the GRB. This is not the case for GRB 020813, where the GRB appears to be illuminating a narrow portion of a shell of matter lying distant to the star.

Can a GRB occur with such a long delay after the supernova that helped produce it? After a decade, it has proved nigh on impossible to produce a model that keeps a neutron star supported against gravity for a few days, not to mention 2 months after it has been formed. Evidence from the work of Donald Burrows suggests that proto-neutron stars may be rather distorted early on as a result of asymmetric accretion from any disc surrounding it within the core of a massive star. These asymmetries rapidly quench as angular momentum is transferred to the star and gravity squashes material into a ball. Consequently, radiation in the form of gravitational waves should eliminate most asymmetry and lead to the stabilization of the young neutron star shortly (minutes) after it has formed. Thus any neutron star born top-heavy should collapse to form a black hole within a few minutes at most – and possibly much sooner. This would remove any possibility that the GRB would occur with such a delay after the supernova.

However, there is a more prosaic possibility. In a marginal case, where a particularly massive neutron star formed, but the shockwave was sufficiently energetic to launch a successful supernova, the majority of the material lying close to the young neutron star will exceed the escape velocity and move outwards from the object. However, if sufficient material remains nearby, it *could* fall back onto the neutron star via an accretion disc and ultimately tip the star over the edge of oblivion. The real question is, could this really take as long as 60 days to happen?

It certainly seems unlikely if not impossible. To achieve this feat the neutron star would have to be poised on the edge of collapse, and the accretion disc would have to be particularly feeble in terms of its actual mass, while extended over a sufficient distance to avoid prompt accretion – or dispersion by any pulsar wind.

So is the supranova of GRB 020813 analogous to standing a pencil on its tip? Perhaps, but there was an unusual precedent set in 1992 – the discovery of the first extrasolar planet. Most readers may feel that this honor fell to 51 Peg b, a hot Jupiter orbiting a Sun-like star. However, this discovery arrived a good 3 years behind the sighting of the first true extrasolar planetary system, orbiting the millisecond pulsar PSR 1257+12.

Revealed by Aleksander Wolszczan and Dale Frail, through the subtle perturbations of the pulsar signal caused by the motion of its planets, this five-body system must have had an unusual origin. Given that it was highly unlikely that planets could have formed so close to a parent O- or B-class star, survived a possible supergiant phase, or the evaporative and dispersive effects of a supernova, they must have formed subsequent to the star's demise. The likeliest origin is in a long-lived (multi-million-year) disc around the pulsar.

Although such a disc may have formed from the debris of a long gone companion, the planets may also have been formed in a disc around the pulsar when it was formed. In support of a long-lived disc, MIT's Deepto Chakrebarty discovered a disc surrounding the magnetar 4U 0142+61 in 2006. Located 13,000 light years from Earth, the magnetar's disc was found using the Spitzer Infrared telescope. Previously evidence has emerged for the presence of discs around the neutron star cousins of magnetars, anomalous X-ray pulsars. In these cases the disc is at least tens of thousands of years old and therefore holds the potential to not only form planets but, in the case of supranova, slowly add mass to the nascent neutron star until it goes over the edge and collapses to form a black hole.

Is there another possibility? Perhaps. Instead of the GRB illuminating the material from its own supernova, perhaps the GRB illuminated the shell of another older supernova remnant, and the high velocity reflected the collision of outward-moving jets of material associated with the GRB with this shell. If the GRB was chronologically linked to a starburst it is certainly not outside the realm of possibility that more than one supernova could have erupted within a suitable time frame that allowed for such an interaction.

Whatever the true situation, this is a problem for the collapsar model, indeed, stellar evolution models as whole. How do you keep an oversized neutron star supported against gravity for a month or more? You probably can't. And as for tweaking the conditions so that fallback is slow enough to prevent prompt collapse, this also seems untenable. Further observations are clearly warranted.

XRFs and Type Ibc Supernovae

X-ray flashes were detected early on by Beppo-SAX, occurring at a rate comparable to gamma ray bursts. There was no indication of isotropic distribution, suggesting a truly cosmological origin. Given the short duration of these bursts (on the order of 10–200 s) the implication was that these softer explosions originated in a similar manner to their more powerful siblings, the gamma ray bursts.

Collapsars, Hypernovae and Long Gamma Ray Bursts 143

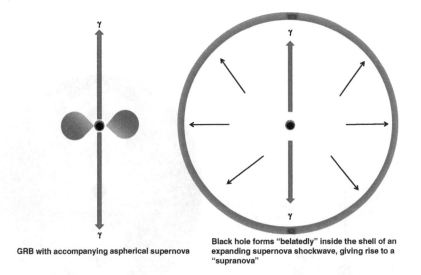

GRB with accompanying aspherical supernova

Black hole forms "belatedly" inside the shell of an expanding supernova shockwave, giving rise to a "supranova"

FIG. 4.3 Contrasting models of a conventional GRB associated with a supernova (*left*) and a "belated" GRB occurring within the pre-existing shell of a supernova – the supranova model (*right*). In the supranova the jets from the GRB illuminate and excite a small portion of the expanding supernova shell, emitting X-rays. The supranova model was suggested as an explanation for the presence of iron-rich debris far from GRB 020813

The bursts detected by Beppo-SAX were so similar to GRBs that it led many to assume that these flashes originated in GRBs that were observed off-axis (Fig. 4.3). Ryo Yamakazi and Takashi Nakamura proposed a model based on earlier modeling work of Kunihito Ioka and Takashi Nakamura. In essence, the GRB erupts as a jet of relativistic material shoots out of the star. Around the axis of the jet, softer, X-ray emission is observed fanning outwards. Based on differences in the viewing angle observers will either catch the GRB if it is seen almost end on, or the softer X-ray emission if the burst is seen at a broader angle.

This model certainly could explain many of the observed XRFs that share burst duration and X-ray luminosities comparable with typical GRBs. However, in the last 10 years, NASA's SWIFT observatory has caught a different flavor of X-ray burst that appears to have a different origin to GRBs (Fig. 4.4).

XRF 060218 was observed on February 18, 2006, by the Burst Alert Telescope (BAT) onboard SWIFT. A short time later SWIFT's optical and ultraviolet cameras captured the afterglow. Follow up observations on the ground by an international team

144   Extreme Explosions

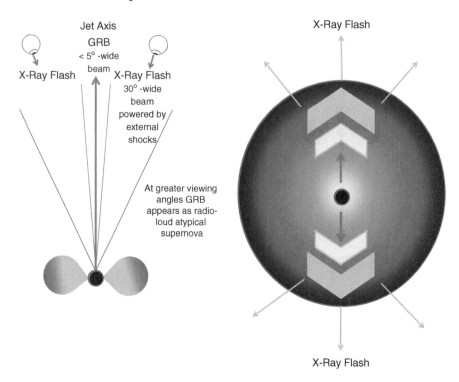

FIG. 4.4 Two models for X-ray flashes. On the *left* a GRB with its narrow jet is observed off-axis, and lower energy emission is produced in broad jets. On the *right* a GRB occurs inside a larger *red* supergiant star. The energy of the jet is absorbed by the extended material of the star, generating slower moving material that emits correspondingly lower energy X-rays

using telescopes in California (Lick) and Chile (European Southern Observatory) confirmed that the burst was associated with a luminous Type Ic supernova.

The term "luminous" needs redefining. A typical (narrow-line) Type Ic event peaks at an absolute magnitude of around −16. The broad-lined Type Ic explosions associated with GRBs, such as SN 1998bw and SN 2003dh, were on the order of 5–6 times more luminous than the narrow-line explosions. SN 2006aj, associated with this XRF, was 2–3 times brighter than a narrow-line explosion, but a few times fainter than the magnitude −19.5 event, SN 1998bw.

Analysis of the X-ray emission suggested that it was unlikely that this was simply a GRB observed off-axis; for one thing, the detected velocities of the ejecta were far slower than those seen in SN 1998bw – perhaps *only* 10,000 km/sec. Moreover, the properties of the progenitor star appeared to differ from those smattering of others that were confirmed hosts to previous gamma ray bursts. Although unobserved, the progenitor appeared to be a small carbon-oxygen star of perhaps only 3–4 solar masses. The parent of this relatively diminutive object may have been only 20 times the mass of the Sun and thus probably achieved its final state through interactions with an unobserved binary companion. It is not expected that a 20 solar mass star will lose both its hydrogen and helium layers without such assistance. Nor is it expected that such a small star could form a black hole following core collapse.

A suitably close binary companion, or one that merged with the progenitor star shortly before its demise, could not only have removed the hydrogen and helium layers but also spun the star up, providing it with sufficient angular momentum to launch a mildly relativistic jet. The question then arises, was the jet launched by a small black hole or by a neutron star? Certainly, manufacturing a black hole would require nearly all the mass if the star fell inwards (fallback), and this was ruled out by observations of a bright and hence nickel-56 rich supernova. We must, therefore, conclude that the progenitor of this XRF was most likely a neutron star and not a black hole. Our range of objects that generate these relativistic phenomena was thus increased.

In the case of SN 2009bb Alicia Soderberg and colleagues detected a very powerful radio burst from this Type Ib supernova. The explosion appeared to be as energetic as SN 1998bw but without the nearly simultaneous detection of an accompanying GRB. Soderberg suggests that the radio emission arose in an analogous manner and environment to radio emission observed in GRBs. However, the GRB was missed simply because the emission was not beamed towards us. Certainly, the energies involved in the radio emission accompanying both this supernova and long gamma ray bursts were analogous. Thus observations with radio telescopes such as the VLA could, in principle, allow the detection of gamma ray bursts that are not detected by conventional gamma ray or X-ray observatories such as SWIFT or HETE-2.

The year earlier Soderberg detected another explosion bearing the hallmarks of a relativistic outburst, but this one was unusual. SN 2008D was widely hailed as the first supernova for which the shock breakout had been observed. However, analysis of the spectrum revealed some more unusual facts that linked it with the GRB phenomena. SN 2008D initially displayed the broad lines characteristic of high outflow velocities (on the order of 30,000 km/sec). The supernova was classed as Type Ic based on an initial absence of hydrogen or helium. However, at later times spectra revealed the presence of helium, and the broad lines, indicative of high velocities, gave way to narrow emission. Was this a Type Ic explosion detonating within a shell of helium ejected earlier by the star? The analogous event would be SN 2006jc. The optical light curve had a double peak with the second peak far brighter than the first – again something that could be explained by an initial explosion followed by a second outburst as the blast wave hit a circumstellar shell.

Paolo Mazzali and co-workers produced an alternative and very successful description of the supernova through computer modeling. Mazzali suggested that the initially broad emission was produced by a low mass of material (0.03 solar mass) moving at the highest speeds, embedded within a larger mass (seven solar masses) of ejecta that was moving at slower speeds. The key to Mazzali's model was how the relativistic jet punched its way through the star and out into space. Mazzali's model starred a progenitor with an initial mass 30 times that of the Sun. After shedding its hydrogen shell through stellar winds, the core completed its evolution and collapsed. The observed X-ray flash, XRF 080109, resulted from a weak jet that managed to escape the star intact, or perhaps more likely a jet that lost much of its kinetic energy as it plunged through the still-massive helium shell surrounding the core (see Fig. 4.3).[1]

SN 2008D was also less than insignificant in terms of its nickel production – at least compared to other gamma ray burst supernova. The explosion synthesized 0.09 solar masses of nickel-56, based on the light curve and maximum luminosity.

---

[1] A new class of ultra-long bursts such as GRB101225A (the "Christmas day burst") seem to be associated with supergiant stars in apparent contradiction of this model for XRFs. Can supergiants generate GRBs after all?

Although somewhat greater than a conventional core-collapse explosion, this was considerably less than the supernovae associated with gamma ray bursts typically found to create 0.5–1.0 solar masses of this element. SN 2008D was not unique. Other supernovae, including SN 2002ap, have been detected that are similarly deficient in nickel-56 but seemingly empowered with relativistically-moving debris, despite lacking any association with GRBs.

So, is there a continuum extending from a GRB associated Type Ic supernova, through to the more common or garden narrow-line Type Ib/c events? Apparently so. SN 2007gr was discovered using the Katzman Automatic Imaging Telescope and was described by Zsolt Paragi. This supernova had only modest overall ejecta velocities, on the order of 6,000 km/sec. However, a small proportion of the mass appeared to be moving at far higher velocities – on the order of 200,000 km/sec, or two-thirds of light speed. Once again, radio emission was the key to its detection. Again, there was no evidence of an accompanying GRB or XRF. Aside from the small mass of relativistic ejecta, the supernova was a fairly typical Type Ic event with no other distinguishing features. Thus, we appear to be descending a ladder from the "hypernova" of SN 2003dh, through lesser mortals such as SN 1998bw, on down through XRF-associated supernovae and into the realm of energetic but otherwise normal-looking explosions. What all these events have in common is a high initial mass, although not extraordinarily so, and a progenitor that has shed its hydrogen and in most cases helium layers to expose its core.

Are there any exceptions? Well, obviously. There are always exceptions that some might say spoil a pretty picture. However, what we're really seeing is a Van Gogh, full of the complexity of interwoven but rapidly delivered brushstrokes. As yet, we've only seen the signature and a small portion of the canvas. Seeing the whole production takes time, patience and often considerable luck. Hinting at what lay beyond these early brushstrokes was SN 1997cy. This Type IIn explosion – one with a clear, abundant hydrogen-rich shell – also showed evidence for excessive kinetic energy, perhaps 3×10^{52} erg or in excess of 10^{45} J. We've already described this as a possible example of a supernova linked to a

GRB; however, it is the presence of hydrogen that sets it apart from the other candidate supernovae.

SN 1997cy was spatially associated with the *short* gamma ray burst GRB 970514. At the time, the supernova was a record breaker in terms of its luminosity, with a peak absolute magnitude of –20.1, or a few times brighter than SN 1998bw. In turn this high luminosity suggested 2.6 solar masses of radioactive nickel had been created in the explosion. Its spectrum revealed some features characteristic of Type Ia (FeII/III lines) and abundant hydrogen characteristic of Type IIs. The H alpha emission is split into broad and narrow features, suggestive of fast-moving shocked material traveling at 30,000 km/sec and a slow-moving (300 km/sec) component that appeared to correspond to flash-ionized, hydrogen-rich material lying around the explosion center.

The light curve was best matched by significant CSM interaction at early stages in the explosion, followed at later times by the simple decay of Co-56. An alternative model was produced by Ken'ichi Nomoto, in which the light curve was synthesized by a simple energetic $(2 \times 10^{45}$ J) explosion that occurred within a dense shell of hydrogen-rich gas. This "hypernova" model had high velocity material violently interacting with the surrounding shell, resulting in the creation of abundant visible light. One key prediction of such a model (and of the basic GRB model) is the formation of abundant radio emission, created as high velocity electrons decelerate and interact within the jet or the shell surrounding the dead star. However, radio emission was not detected, suggesting a lack of a relativistic outflow, in flat contradiction of both scenarios.

If we accept that SN 1997cy was a Type IIn event analogous to SN 2006gy or 2006tf, then we would still expect detectable radio emission. However, the observations weren't taken until 16 months after the explosion (at 1.384 and 2.496 GHz). It remained possible that these were simply too insensitive or even too late to allow the detection of emission from a supernova at the redshift of 0.063. Therefore, despite this explosion's potential to be the only Type IIn event linked to a GRB, this has remained a cold case ever since. In 1998 Lisa German and colleagues noted that Lifan Wang and Craig Wheeler had already provided a statistical *hint* that GRBs were associated with Type Ic events. There were a small number of such associations. However, the limited ability to pinpoint GRBs with suitable accuracy meant that proving such a correlation was weak at best.

SN 2010jp: The First Jet-Powered Type II Supernova

If SN 1997cy failed to link hydrogen-rich Type II supernovae with collapsars, then the Palomar Transient Factory's PTF 10aaxi, also known as SN 2010jp, completed the circle.

This was an unusual explosion in many regards. The spectra indicated material expanding away from the explosion with very different energies in different directions. It was low in overall luminosity, with an absolute magnitude peaking at only −15.9. This is far below other GRB supernovae or the Type IIn supernovae described in Chaps. 3 and 8. The low luminosity also suggested that very little radioactive nickel-56 was synthesized in the explosion, perhaps only 0.003 solar masses. This is a twentieth the amount typical of other Type II explosions – or less than a hundredth of the amount produced in GRB-supernovae, such as SN 1998bw. This was clearly not an impressive supernova. However, the peculiar velocities of the ejecta were remarkable.

Nathan Smith (University of Arizona) and co-workers describe a supernova with one hydrogen-rich component moving at a slow 800 km/sec, a high velocity component indicating material moving at more than 12,600 km/sec and finally a third, very broad component indicating velocities of 22,000–25,000 km/sec. What was the origin of these disparate velocities?

Smith and colleagues suggest that the slowest-moving material corresponds to hydrogen-rich gas lying around the explosion center, which was shock-heated by the blast wave of the supernova. The outgoing shock had smashed into dense circumstellar material and decelerated to 800 km/sec. Nifty, but slow compared to the other material that was detected. The velocity of this material seemed rather high to be the stellar wind from a red supergiant progenitor. Instead, it was mostly likely generated by an eruption on a star similar but lower in mass than a luminous blue variable (see Chap. 3). The inferred mass of the progenitor was most likely around 25–30 solar masses, putting it likely above the maximum mass for a red supergiant star but still within the domain of conventional (yellow or blue) supergiants.

The fast-moving material appeared as intriguing blue and red "spectral bumps" that suggested these were associated with hydrogen-rich gas moving in opposing directions at very high speed (over 20,000 km/sec). Material directed towards us produced blue-shifted features in the spectra, while material receding from us produced the red bumps. These spectral features persisted throughout the period of observation, clearly linking them to the supernova and not some underlying association.

Supernova SN 2010jp was thus the first Type II supernova to be associated with underlying jets of material. There was already a suggestion that supernovae would be aspherical, and indeed this is seen in supernovae such as SN 1987A. However, the degree of asphericity in Type II supernovae is rather modest at best. More compact progenitors, such as the Wolf-Rayet stars that give rise to GRBs, clearly display strong asphericity. Indeed, this is the underlying feature of a GRB-supernova. It is powered by relativistic jets of material. The presumption is that these jets become ameliorated by their passage through the star. As such, very little asphericity is likely to persist in the explosion if the jet has had to punch through extensive layers of hydrogen and helium before reaching interstellar space.

So, why was SN 2010jp different? The authors suggest that this was linked to the metal content of the star. SN 2010jp is associated with a very low metallicity dwarf galaxy, or a distant section of an interacting pair of galaxies. As such the metal content of the progenitor was likely very low. Indeed, analysis of the spectrum of the explosion indicates a low metal content in the outer layers of the star. Why is this important? Low metallicity is expected to result in a more compact star, such that the hydrogen-rich outer layers lie in a denser configuration, closer to the core of the star. When the core imploded the jets that were produced subsequently had less far to travel to reach the out-

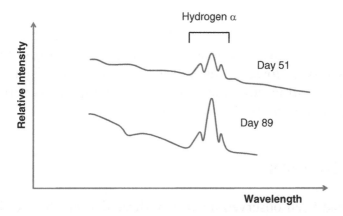

Fig. 4.5 The spectrum of SN 2010jp showing the unusual triple peaked H-α feature attributed to opposing jets of material. One peak represents a blue-shifted emission with the other red-shifted, as it moves toward or away from us, respectively

side of the star. Less energy was dissipated en route, and the jet structure persisted to be observed some time later. However, the compact nature of the star ensured that much of the energy in the jets was soaked up, blowing the envelope outward. This left little energy to drive the resulting supernova, which was correspondingly faint (Fig. 4.5).

Finally, the slow-moving spectral component may have been produced in a small LBV-like eruption prior to the supernova. This explosion shed much of the star's outer hydrogen-rich envelope, allowing the jet to penetrate the star without much loss of energy.

Collapsar models of Andrew MacFadyan, Stan Woosley and Alex Heger predicted that such low metallicity stars might indeed produce aspherical explosions following the birth of a black hole at their heart. The low metallicity produces the compact star and thus preserves the jet structure of the explosion. Indeed, these same models suggest that without the jets, the black hole will simply and rather quietly consume the compact, denser star.

What we can say now is that the majority of supernovae that are associated with either GRBs, or the less energetic XRFs are clearly Type Ic events. Not only does this provide a compelling consistency, these explosions are readily explained within the framework of the collapsar model. Furthermore, as we have seen, there are good reasons to doubt that GRBs can be associated with bulkier and often more massive hydrogen-rich progenitor stars such as LBVs. However, as SN 2010jp shows, you can still get rather interesting outcomes when jets attempt to penetrate the envelope of a more bulky hydrogen-rich star.

Conclusions

GRBs had been observed for nearly three decades before the first of a slew of fading after-glows was bagged by Beppo-SAX in 1997. The ensuing revolution saw the confirmation and elaboration of the collapsar model proposed by Stan Woosley at the start of the 1990s. As more data streamed in we saw the linkup between certain energetic Type Ic explosions and GRBs and the discovery of a mysterious series of dark bursts that seemed to have no causative link to any supernova. Moreover, there were a handful of events that seemed to follow, not accompany, supernovae. These supranovae are still a challenge to theory.

Lying beyond the signature on the canvas are the weaker but schematically linked explosions known as X-ray flashes, or XRFs. At present some appear to be weaker analogs of GRBs, while others may simply be GRBs observed off-axis – seen at an angle outside the jet that produces the gamma ray emission. Further down the line are energetic Type Ic explosions that produce neither X-ray nor gamma ray emissions. These relativistic bursts link the standard narrow- and intermediate-line explosions with the broad-line burst-associated events. Thus, after a decade or so since the first associations were made between GRBs and supernovae, we have a family of events associated with the deaths of Wolf-Rayet stars. What remains to be elucidated is the factor, or factors, that divide a typical Type Ic explosion from its more energetic brethren.

5. Death by Fallback

Introduction

We have already seen that some supernovae can falter when large amounts of the shocked outgoing material achieve insufficient velocities to escape the developing neutron star. However, can this be so extreme that the supernova fails completely and the star dies alone in the dark?

One imagines that as a star ascends the main sequence, reaching greater and greater masses, that the bang such a star delivers when it expires increases in turn. This might be a simplistic assumption; after all, more star more energy. We already know that a star with 20 times the mass of the Sun dies more violently than smaller stars. However, nature once again seeks to confound our expectations.

Imagine a star 50 or 60 times the mass of the Sun. Silicon has been furiously fusing to make a massive core of iron over the preceding hours. After less than 24 h, the core is filled with iron – over one and a half Sun's worth of metal is squeezed into a sphere no larger than Earth. Bereft of any possibility of generating further energy, the core promptly implodes, shrinking to something the size of Earth in less than a second.

Like its smaller cousins further down the main sequence, this monster star belches forth a storm of neutrinos that hurtle unimpinged through the star, heralding its destruction. A neutron star has formed, and a shockwave begins to move outwards into the massive, overlying envelopes of silicon, neon and oxygen.

Like a lone spectator trying to fight his way out of the stadium before the start of a soccer match, the outgoing shock falters against the opposing tide, then stalls. Soon all outward momentum is lost, and the massive outer parts of the core – particularly those not supported by rotation – begin free-falling inwards. What will the outcome be?

We've already seen that some stars appear to form black holes, announcing their demise with a gamma ray burst. The star is shredded from the inside out by a narrow jet of radiation beamed from the nascent black hole. In some cases, at least, this beam is accompanied by a Type Ic supernova. However, is this always the case? Can a black hole form in the dark? Can a black hole consume its parent star without so much as a cosmic burp? And perhaps most pertinently, can we ever hope to observe something that is inherently dim?

The Mystery of Cygnus X-1

This is the question posed by the X-ray binary system CygnusX-1. Cygnus X-1 is a peculiar pairing of a 10 solar mass black hole and an 18 solar mass blue supergiant star. The latter is slowly being eaten by its dark companion. In the process, some of the matter, stripped from the companion, spirals down the gravity well of the black hole. In doing so friction heats it up to the point that it emits a torrent of X-rays. These are visible from Earth. It is these X-rays that give the system its designation "Cygnus X-1," for the first detected X-ray source in the constellation of Cygnus. However, the system doesn't just emit X-rays. Black holes are often messy eaters, and unsurprisingly some of the stolen matter is also ejected in radio-emitting relativistic jets, as well as visible and other forms of electromagnetic radiation.

As Félix Mirabel and Irapuan Rodrigues discussed in a 2003 article in the journal *Nature*, given the mass of the companion star and the other stars in the surrounding stellar association, Cygnus X-1 should be less than 10 million years old. A best estimate puts the system's age at no more than 7 million years. This implies that the black hole would have formed less than 1 million years ago. So why is there no supernova remnant detectable in the vicinity of the black hole? After all, although a few hundred thousand years old, the supernova blast wave would have heated any surrounding gases. This should have generated a ragged shell of hot material that profusely emitted X-rays. Yet nothing is detectable. There is no evidence that a supernova ever took place.

It is reasonable to assume that the star that left a 10 solar mass black hole began much larger – perhaps more than 40 solar masses? Even if it sheds the bulk of its mass in a Wolf-Rayet stage before dying, the star must have been massive. Intuitively, the star must have ejected the majority of its leftover mass in a supernova. Yet, the best estimates suggest that if matter was ejected it must have been less than 1 solar mass.

Furthermore, it is apparent from observations of the motion of some neutron stars and black holes that when they were formed the supernova gave them a nascent kick. This throws them far from their birth places at high velocity. Yet, when Mirabel and Rodrigues look at Cygnus X-1, the binary system is still clearly part of the stellar association with which they were formed – they've gone nowhere.

When one looks at the black hole binary system GRO J1655-40 the contrast with Cygnus X-1 couldn't be greater. Here, the black hole system is speeding away from its association, implying a traumatic birth in a supernova. So, if some black holes show evidence for their birth in a supernova, why do some not? Moreover, for those that do not, what happened when the star died?

Clues to the formation of Cygnus X-1 come from modeling work by Roger Chevalier, Stan Woosley and colleagues. In Woosley's models some supernovae, essentially, fail to thrive. The ultimate source of energy driving the explosion comes from the imploding core (Chap. 2). Neutrinos and copious heat are released as the core collapses. However, where the core implodes directly into a black hole, or where there exists excessive mass around the developing neutron star, the outgoing shockwave has insufficient energy to drive material outward. A reverse shock front, akin to a stopper in a set of rapids, develops. This stalls outward moving material and allows it to efficiently fall backwards and be consumed by the growing, dark heart of the star.

If sufficient material is accreted the neutron star exceeds its upper mass limit. The limit lies somewhere in the region of three times the mass of the Sun. Neutron degeneracy, the crutches supporting the star against further collapse, crack and fail. The formation of a black hole is now an inevitable consequence. Under the correct conditions, perhaps where the star is slowly rotating or where there is an insufficiently strong magnetic field, almost all

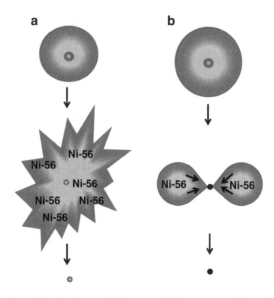

Fig. 5.1 A death in the dark. In (a) the star blows up and nickel-56 enriched material is blown out into space, powering a strong supernova. In (b) the majority (or perhaps all) of the nickel-56 enriched material created in the core collapse is sucked down into the growing *black hole* and is not available to power a supernova. The star dies alone in the dark

of the star plunges back into the yawning abyss. Little star stuff evades the growing monster, either in the form of jets or as a supernova (Fig. 5.1).

Although there is considerable uncertainty in the precise mechanism that leads to black hole formation, there is now observational evidence that at least some black holes have formed recently without much show and tell. Notably, three papers appeared back-to-back in the journal *Nature* in 2006, describing the unusual characteristics of a GRB detected on June 14, 2006. The GRB had a duration typical of long gamma ray bursts (102 s). Unlike many eruptions that had been detected in the previous decade, this one lacked the telltale signs of an underlying Type Ib/Ic supernova. GRB 060614 was very similar to another burst detected the year before, implying that a small but significant number of long duration GRBs could have an origin distinct from the Type Ic associated events.

GRB 060614 had the characteristic soft spectra indicative of longer bursts, but it also had some unusual features. The first 5 s of the June 14, 2006, burst was dominated by hard (>100 keV) gamma rays, but these were followed by a much softer 100 s phase with declining overall intensity. Despite the long duration, the burst pattern was very reminiscent of the harder, short-duration bursts detected by the SWIFT probe.

The prevailing and highly successful model for short duration gamma ray bursts involves the merger of two neutron stars in a tight binary system, or the merger of a neutron star and black hole. As the neutron stars approach one another violent tides begin to stretch and distort both stars until iron- and neutron-rich matter coalesces into a temporary massive object, surrounded by a very hot neutron-rich disc of matter.

There is a possibility that such mergers are the site of synthesis of the heaviest r-process elements such as gold and uranium, rather than core-collapse supernovae, but at present this remains unproven. Either way, the disc surrounding the coalescing stars doesn't last long. Soon after the merger commences, the very hot object that has formed becomes unstable and collapses under its own weight. The outcome is a black hole with a mass approximately three times that of the Sun. Such black holes rapidly consume the remains of the two stars that continue to swirl in a disc around the black hole.[1]

In a manner presumably analogous to that seen around the more massive holes produced by collapsars, jets are launched from the vicinity of the black hole, draining away the large quantity of angular momentum present in the disc. These opposing jets generate the GRB. As the disc around these black holes is also relatively small in terms of its mass, the process lasts less than 2 s, and the burst is thus short in duration. Finally, with the absence of matter surrounding the disc (in the form of a star), the energy of the radiation generated by the jets isn't attenuated, and the burst is harder –that is, has shorter and more energetic gamma rays than long duration bursts. Consequently, these bursts should be rather distinctive compared to the long GRBs associated with massive stars.

Moreover, such mergers should happen some considerable time after star formation has ceased in the area in which the burst occurs, simply because such mergers unfold over tens or hundreds of millions of years after the system was born. This gives the

---

[1] The accompanying weak predicted explosions have now been detected and are known as "Kilonovas" (Nature, 8th August 2013).

binary system enough time to leave its dispersing stellar cluster or association. Yet, spectral analysis of the region in which the burst of June 14, 2006, occurred indicated that it was a star-forming region, displaying prominent H-α emission, typically associated with the presence of young, hot blue stars.

Therefore there is a dichotomy in these results. GRB 060614 had the duration and point of origin typical of long duration GRBs but the spectrum of a short GRB. How can these results be reconciled?

Close scrutiny of the spectrum of long and short bursts led Grehels and colleagues to suggest a new GRB classification scheme based on the discovery of the dark but long burst. The early schism between the two types of bursts was based solely on the length of the burst itself. The short bursts lasted less than 0.2 s, and the long bursts, well, longer than this. In essence this is a biologist separating animals based solely on diet. It's a little crude and in need of some enrichment.

Indeed, there was ample evidence of spectral features that could be used to differentiate bursts, but hitherto had not really been used. Grehels suggested classifying bursts based not on the duration of the burst but on a feature called a spectral lag. Long bursts tend to have a prompt hard portion lasting typically less than 5 s. This is followed by a short gap (milliseconds to seconds in length) with little or no emission, and terminated by a longer soft phase of gamma ray emission lasting up to 3 min or so. The lag in GRB 060614 was 366 milliseconds – well within the range of short duration GRBs.

Subsequently, as well as having an overall, shorter burst, short GRBs also have a much narrower gap between the two portions of the burst. Thus, in terms of overall length, the 102-s GRB was clearly long, but the gap between the prompt, hard portion of the burst, and the later softer emission, fitted better with the short duration camp of explosions. Analysis of the HETE-2 and SWIFT databases revealed further anomalous bursts that had features characteristic of both types of GRB, and the suggestion was that this was an underestimate, as more distant and hence fainter GRBs could be missed.

GRB 060614 had no associated supernova, indicating little, if any, radioactive nickel-56 escaped the developing black hole. Therefore, what sort of object created this burst, and the other bursts like it? Moreover, did the dark burst give an indication as to whether the black hole in the Cygnus X-1 system was formed in the same way?

GRB 060614 might best be explained by the death of a massive Wolf-Rayet star, where the majority of the star falls straight into the black hole. Should insufficient energy be provided to the shockwave, the radioactive nickel synthesized around the hole will simply fall into it and be unavailable to power any sort of supernova. The radioactive decay of nickel-56 powers the declining light curve of supernovae. Since most of this is formed as the shockwave powers through the silicon shell and into the oxygen burning shells, very little of this escapes the maw of the black hole (less than 0.01 M). The majority is swallowed almost as soon as it is created. Hence the supernova, if one is seen at all, is very weak. Observations of GRB 060614 constrained the luminosity of any accompanying supernova to less than an absolute magnitude of −13.5, or 100 times weaker than a conventional core-collapse Type II supernova.

Consequently, the death will happen in the dark. This occurs when the progenitor is particularly dense and there is little mixing in the explosion. The shockwave falters early on, and the star simply cascades inward. The only evidence of its demise is a burst of gamma rays and X-rays and the light going out. The collapsar model remains intact, and the star blinks out of our universe.

So, how can we link this fate to the presence of the relatively massive black hole of Cygnus X-1? Cast our minds back to our massive, dying star – the progenitor of the black hole in the system. As the iron core reached the Chandrasekhar limit it imploded. The neutrinos ripped outward, with the developing shock wave stalled under the weight of the in-falling oxygen-rich shell above. The proto-neutron star's magnetic fields whipped outward and generated a jet. Under the crushing weight of the oxygen shell, the core then continued its descent into oblivion, forming a black hole. The surging jet was then reinvigorated by the additional gravitational pull of the growing black hole, which drew the accretion disc in at an even greater pace.

Within seconds the jet had punched through the star, continuing outwards until millions of years later, some of the gamma rays forged within it impinge on our detectors in orbit around Earth.

The jet would have imparted enough oomph to the surrounding stellar material to bow it outward, but not enough to drive most of it to escape velocity. Instead the material would have been pancaked into a swirling, thick accretion disc, and as this spiraled down the plughole of the newly formed black hole a second wave of softer gamma ray emission would have followed. In this scenario the hard part of the burst is associated with the initial outgoing jet, and the longer, softer part of the pulse is associated with the burps and coughs as the accretion disc feeding the black hole drains away. A small fraction of the material feeding the hole is trapped by the system's magnetic field lines and whipped outward, forming the tale in the gamma ray burst.

Controversial Supernovae

However, can we have a supernova without a GRB in this scenario? It is already clear that the majority of Type Ic supernovae occur without accompanying GRBs. If there was no GRB what would the supernova look like if it was launched successfully? An interesting example of such a supernova might be SN 2008ha. This supernova is generally regarded as a low luminosity thermonuclear event (Chap. 13). However, following careful analysis of its spectra by Stefano Valenti and others (University of Belfast) presented a persuasive argument for SN 2008ha having an origin in the formation of a black hole. Valenti's case was based on a number of unusual features in the spectrum, suggesting that this explosion originated in a massive star that had shed most or all of its hydrogen-rich outer layers in strong stellar winds.

SN 2008ha had at most 5×10^{49} erg (5×10^{42} J) in visible energy, making it two or three orders of magnitude weaker than GRB-associated supernovae. Could this weak supernova be a dark (or at least dim) event without an associated GRB? Do some black hole-forming GRBs occur without supernovae, while other black hole birthing supernovae occur without a GRB? The chemistry of the ejecta certainly fitted better with the predictions for an unusual shell supernova (Chap. 13), but perhaps our understanding is wrong.

The spectrum of SN 2008ha appeared to match Type Ic events overall, suggesting that there was a detonation within a Wolf-Rayet star. There was no evidence for either hydrogen or helium in its spectrum, nor the silicon line characteristic of Type Ia events. Although helium is not seen in Type Ia events it can be present as a weak feature in Type Ic explosions. Its absence here cannot therefore be used to discriminate between the two classes of explosion. Further scrutiny of the spectrum revealed narrow emission lines characteristic of relatively slow-moving ejecta, and this coupled to the low overall energy suggested a radioactive nickel yield considerably less than 0.1 solar masses – perhaps as little as 0.01. Indeed, the yield was probably an order of magnitude less than a conventional core-collapse supernova.

As Chap. 13 goes on to describe, SN 2008ha was clearly enigmatic. Although the suggestion that SN 2008ha was a core-collapse event is controversial, it cannot be completely ruled out. Moreover, in terms of energy and overall luminosity, some observations of SN 2008ha allowed Valenti to suggest a kinship of this event with a small but recognizable group of faint Type II-P events. There was the suggestion that these weak Type II-P explosions may share a similar origin – perhaps the formation of a stellar mass black hole.

The leading possibility for such faint explosions is that core collapse leads to the consumption of the inner portions of the star. With little radioactive nickel escaping the black hole, the supernova was dim and had only low velocity ejecta. Unlike the more spectacular GRB-associated events, perhaps the progenitors of these explosions, whether Type II-P or Type Ic, are spinning slowly or contain intrinsically weak magnetic fields. Consequently, the jets that are characteristic of GRBs are not launched, and the explosion is weak.

SN 2008ha is not without precedent or antecedent. An entire class of unusual supernovae, SN 2002cx-class events, may be related. These, again, are generally thought of as peculiar Type Ia (thermonuclear) supernovae (Chap. 13). However, instead they could be members of a class of low luminosity, core-collapse explosions that owe their origins to the formation of a black hole. Valenti and colleagues certainly don't come down on one side, hedging their bets that this, and other, weak explosions might equally mark the death of stars through electron capture (Chap. 7). If nothing else the circumstances surrounding SN 2008ha certainly reveal a healthy level of debate and speculation within the astrophysical community. This is an attitude that should be welcomed.

Populations

In 2004 Andrea Pastorello and colleagues proposed that 4–5 % of all Type II supernovae would be associated with such core-collapse scenarios. Observations of GRB 060614 fit these predictions and suggest that some black holes are indeed formed in the twilight, if not entirely in the dark. As Félix Mirabel and Irapuan Rodrigues discussed a year earlier, the Cygnus X-1 system may be the outcome of one of these events.

There are other peculiar, faint Type II explosions, such as SN 1994N, SN 1997D and SN 1999br and SN 2001dc. The low luminosity of this group of explosions and the low mass of nickel-56 ejected (less than 0.01 solar masses) clearly sets them aside from more typical Type II events and links them to events such as SN 2008ha. Analysis of the light curve of SN 1997D also lacked evidence for the kind of energy input into the expanding supernova remnant that you would expect from a nascent pulsar or magnetar (10^{43-44} J). This led Pastorello and others to conclude that these events marked the formation of a stellar mass black hole by fallback. The progenitor in each case was not observed, but the implication would be that a massive star hosted these explosions – each one weighing in at 25–40 solar masses.

Indeed, evidence has emerged for an entirely new class of intrinsically dark GRBs, epitomized by GRB060614 and GRB 010921.

Can we explain these faint supernovae through an alternative mechanism? Examination of the most likely alternative scenario also makes for some testable predictions. Accretion-induced collapse (A. I. C.) supernovae (Chap. 7) would be expected to be faint and potentially hydrogen- (and possibly helium-) deficient. Electron-capture events occurring within super-asymptotic giant branch (SAGB) stars might appear as faint Type IIn supernovae. Even if intrinsically weak, such explosions would inevitably encounter the dense SAGB wind ejected in the millennia prior to core collapse and generate the narrow emission lines characteristic of collisions between the blast wave and circumstellar material. In principle, these differences should be apparent in the early and late-time spectra, and the overall photometry of the explosion. Similarly, as discussed in Chap. 7, there are some predicted and very distinctive chemical signatures in the ejecta of electron-capture supernovae that should make them distinguishable from weak Type Ic explosions.

The electron-capture-induced collapse of the core of an 8–10 solar mass star appeared to be ruled out in the case of SN 1997D by the absence of a neutron star remnant. However, it took some time for the neutron star to be identified within the nearby and very prominent Cassiopeia A remnant. Therefore, just because we cannot see a neutron star does not automatically mean it isn't there. Indeed, there could be a newly minted black hole, but there could also be too much background radiation to allow the detection of a young, hot stellar remnant.

We should reflect upon the dictum of Carl Sagan – absence of evidence does not equate to the evidence of absence. Further observations are clearly warranted. However, it is possible, perhaps even probable, that a sizable fraction of the observed faint supernovae are fallback events, rather than electron-capture supernovae. Moreover, should this scenario prove to be true, we will then have a handle on these explosions that we can use to lever out other such events from more generic core-collapse supernovae.

Conclusions

Although sporadic and rare, there are increasing numbers of observed explosions that are unusually weak. The question is, what is the reason for this paucity of energy in the explosion? It might be that some are electron-capture events, while others mark the formation of a stellar-mass black hole without a GRB. Other GRBs appear to be intrinsically dark. What happened to the star when these GRBs were launched? Although there are likely to be several answers to explain the intrinsic faintness of some explosions, the formation of stellar mass black holes is likely to lie behind a proportion of them. How big this proportion is, at present, anyone's guess. The breakthrough in proving the link between "dark GRBs" and core-collapse supernova will probably come with improvements to the optical systems designed to detect potentially weak explosions, particularly those over long distances.

Distant supernovae – the bright ones – are recorded at present with a handful of pixels. Considerable advances will be needed if we are to observe faint supernovae underlying a gamma ray burst. And, of course, it is always possible there is no underlying supernova with some of these bursts. All the material that escapes the path of the jet may swirl down the plughole and out of our universe. We will then be left with an unusual continuum of faint, peculiar supernovae, from the truly invisible, GRB-associated gasps, through to detectable but still under-luminous explosions, such as SN 1997D. Discrimination between this and the electron-capture models then falls to spectroscopy of the expanding debris and to the detection (or absence of detection) of a neutron star or black hole remnant.

The peculiar supernovae, described in Chap. 13, are almost certainly going to be explained primarily by some form of thermonuclear event. However, we should not rule out alternative explanations until more evidence is available.

6. The Formation of Massive Stars by Collision and Their Fate

Introduction

Prior to the early part of the last decade, there were two rival theories for the formation of massive stars. In one model massive stars were produced by the merger of lower mass protostars within dense clusters of nascent stars. In the second model massive stars formed in exactly the same way as their lower mass brethren – through steady accretion of matter onto a dense protostellar core through an accretion disc. It was presumed that the former was the more likely model, as radiation pressure from the developing star would act to drive material away from it. In Chap. 2 we discussed the identification of massive protostars surrounded by thick, massive accretion discs. These observations have made a very solid case for the second accretion model.

However, despite direct observational success it doesn't eliminate the merger model entirely. Indirectly, the merger model appears to work well when explaining some unusual features of massive star clusters seen in starburst galaxies. It is to these star clusters, and to some unusually massive stars, that proponents of stellar mergers turn. Moreover, computer simulations appear to suggest that under the correct circumstances stellar mergers are an inevitable feature of dense star clusters. These observations and models are what this chapter focuses on. The outcomes can be dramatic and, on occasion, thoroughly odd. Although many of the conclusions have yet to be proven, the suggestions are so dramatic as to make them worthy of special focus.

A Lack of Interpersonal Skills: Harassment and Scandal

In some circumstances, particularly massive molecular clouds can be compressed and instigate star formation. This may be a consequence of galactic mergers or through the less extreme process of galactic harassment. Here, a passing galaxy's gravitational tug sends massive clouds of gas colliding and collapsing into its neighbor. M82 is a particularly dramatic and nearby example, having been harassed by the more massive M81. In either of these situations, particularly, massive clusters of stars containing tens of thousands of members can be formed. Examples of these include the Large Magellanic Cloud's R136 cluster, or NGC 4244 in M33. Both of these galaxies have been harassed by larger neighbors – the Milky Way and Andromeda (M31), respectively.

In 2002 Jarrod Hurley and Michael Shara (The Department of Astrophysics at the American Museum of Natural History) published an enlightening paper on the seemingly scandalous behavior of stars in globular clusters. These ancient, seemingly passive balls of hundreds of thousands of stars hang out with unusual pairings of objects and disturbingly youthful-looking beasts that defy initial classification.

Using the GRAPE-6 supercomputer, the orbits and fates of 20,000 stars, including hundreds of binary systems, were followed over a time period equivalent to billions of years. Shara and Hurley anthropomorphically described the nature of some of these stars as promiscuous, which was perhaps something of an understatement. In the course of evolution one star ended up six times heavier than the main sequence turn off, substantially altering its evolution. Another binary ended up as a neutron star helium giant pairing, which in turn ended up merging. The resulting object formed within the first 94 million years of the cluster's life and was a very curious beast indeed.

Merger of a neutron star with a giant star is theoretically possible. Although a complete understanding of the events associated with the merger needs clarification, the outcome may be a so-called Thorne-Zytkov object – a giant star with a neutron star core. Thorne-Zytkov objects, named after the two astrophysicists that proposed them in 1977, would show some unusual and characteristic surface chemistry. They would also be incredibly unstable.

As the neutron star accretes hydrogen or helium onto its surface, the material burns in a series of reactions called the rapid-proton or *rp*-process. These reactions produce a characteristic set of proton-rich isotopes of intermediate mass elements. Vigorous convection within the overlying envelope of the star would dredge these up onto the stellar surface. These odd elements would appear over time at the surface of the giant star.

Thorne and Zytkov proposed that, at least in some instances, the inevitable outcome of such a merger would be a stellar mass black hole. The neutron star would eventually accrete sufficient material to trigger its collapse to form a black hole. This in turn could herald its formation as a long gamma ray burst. However, it is also possible that in the process of orbital migration, the spiraling motion of the neutron star would disperse the outer layers of the giant in a common-envelope phase, leaving a tight binary instead (Fig. 6.1).

Regardless of the outcome of this particular example, what is apparent from Jarrod and Hurley's analysis is that stellar mergers are common in dense star clusters. Of the 20,000 objects in one study 500 mergers were observed over the 5-billion-year run.

Interesting though this is, these models simulated a period of billions of years – far in excess of the life of a massive star. So what's the connection?

It is likely that when a star cluster is young there are far more opportunities for stellar collisions. In the older clusters, studied here by Jarrod and Shara, most of the remaining stars are relatively lightweight. All the initial massive stars having long since passed away. However, the effect of massive stars on a dense star cluster extends beyond the life of the star itself.

As massive stars age and die their powerful winds and subsequent supernovae drive vast amounts of mass from the stellar cluster. This in turn loosens the gravitational bonds between the stars, making collisions less likely. By inference collisions are more likely early on, when massive stars may be abundant. Moreover, massive stars are larger bodies and form a greater cross-sectional area through which collisions can occur. This also increases the likelihood of a collision between stars.

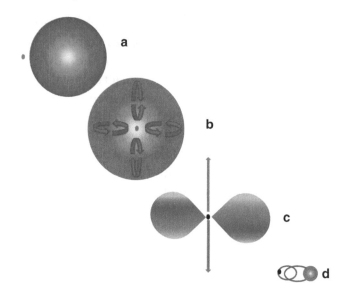

Fig. 6.1 The formation and fate of a Thorne-Zytkov object (TZO). In (**a**) a neutron star is engulfed by an expanding red supergiant. The neutron star spirals into the helium core of the supergiant (**b**). The resulting object is a TZO. The rp-process begins to churn out exotic, proton-rich isotopes of various elements. In (**c**), accretion of matter onto the neutron star causes it to implode and form a black hole. This may be associated with a long GRB. What remains of the envelope may condense to form either planets or low mass stars that tightly orbit the black hole (**d**). It may well be that the red supergiant would be dispersed as the neutron star spirals in. However, the presence of some rare, unusual pairings of low mass stars and black holes suggests that in some instances a TZO may be an intermediate phase in the formation of these systems

Understandably, the expectation is that in dense, young and massive clusters there are ample opportunities for collisions and mergers of massive stars. The outcome of these events would be the formation of ultra-massive objects, through which rather aberrant evolutionary paths are possible. Most importantly, even when mass loss is furiously whittling away at the envelopes of these stars, a sufficiently high rate of stellar collision can keep the masses of the largest objects topped up, right on through until core collapse.

In 1999 Simon Portegies-Zwart and colleagues published simulations based upon the characteristics of the R136 cluster in the Large Magellanic Cloud. Although these simulations neglected the effect of binary systems, which incidentally tend to promote interactions, the researchers found that after only 3–4 million years of simulated evolution a central star was formed through successive collisions. The mass of this object exceeded 100 solar masses.

Observations of R136 suggest that there are a number of unusual very massive stars with spectra suggesting that they are WN stars. WN stars are nitrogen-rich Wolf-Rayet objects that still contain a substantial amount of hydrogen in their envelopes (Chap. 3). These stars are in the process of casting the outer, hydrogen-rich layer off, exposing their helium cores. The nitrogen is forged by the CNO cycle (Chap. 1).

When hydrogen is plentiful nitrogen is recycled into oxygen and back into carbon to repeat the cyclic helium-synthesizing process. But as hydrogen begins to run out, nitrogen accumulates faster than it can be destroyed, and the O-class star begins to morph into a Wolf-Rayet star in an intermediate stage.

The spectra of these objects are intriguing, showing contradictory signs of both age and youth. Their spectra contain violet edges, indicative of later, more evolved stars, yet these stars appear otherwise youthful. This observation, combined with their hydrogen-rich composition and their unusual brightness, suggests that they may be the products of collisions between hydrogen-rich stars and more massive, more evolved objects, perhaps LBVs or WR stars.

Observations made of R136 in 2010 suggested there may be a star with a mass as great as 260 times that of the Sun. Although this interpretation may be tempered by further higher resolution observations, there is a clear suggestion that there may be a monster star buried within the light of countless others. The mass of this object may well be in excess of theoretical predictions that focus on star formation by accretion alone. If true the most likely scenario is that it formed through a process of runaway mergers between already massive stars. Despite this object's propensity to lose weight, its attempts may be thwarted by further mergers before its core evolves and becomes unstable.

Can observations reveal any more exotica in star clusters? Well, perhaps. A nearby galaxy, M82, is currently undergoing a starburst as a result of a recent period of galactic harassment by its near neighbor M81. Several million years ago M81 swept past its smaller companion and instigated a period of turmoil.

Large clouds of gas in M82 began colliding and collapsing, generating a massive wave of star formation. Within the overall mêlée of star birth a few massive clusters formed that bear all the hallmarks of young globular clusters. Star numbers within these clusters exceed 100,000, and a large fraction of these newborns are massive O- or B-class objects. Characteristically, the diameter of these clusters is small, less than 6 light years, and the stellar density exceeds a million times that seen in our neighborhood.

One of these clusters, MGG 11, has an unusual feature – intense X-ray emission. The star cluster is particularly massive, weighing in at over 350,000 solar masses. Intriguingly, a more massive and nearby cluster, MGG 9, does not. What is it about MGG 11 that might explain the anomalous X-ray emission?

Analysis of MGG 11 indicates that any suitably massive star could sink to the cluster's core before it had time to evolve off the main sequence. In 2002 Simon Portegies-Zwart and colleagues proposed that the propensity of stars within MGG 11 to quickly find their way to the cluster core could have resulted in multiple collisions involving stars 30–50 times the mass of the Sun. After as many as 100 collisions, the formation of a particularly massive star resulted – one weighing in at 800 or more solar masses.

Although only 3–4 million years old at inception, core collapse would quickly follow this altered star's birth, forming a black hole with at least 350 times the mass of the Sun. Objects like this are named intermediate mass black holes (or IMBHs for short), intermediate in mass because they are perhaps 30 or more times more massive than the black holes formed from single massive stars, but considerably less than the million or more solar mass objects found at the heart of all large galaxies.

IMBHs are predicted to form early in the life of all massive galaxies when star formation is rife and multiple dense clusters of stars are formed. These ancestral black holes then merge to form the super-massive younger siblings that lie in wait at galactic centers.

MGG 9 seems to have escaped this fate, despite its greater mass. Why? Simon Portegies-Zwart's analysis suggests that although this cluster is more massive than MGG 11, its lower overall stellar density has prevented the required number of rapid, successive collisions needed to fuse stars together into the dark monster seen

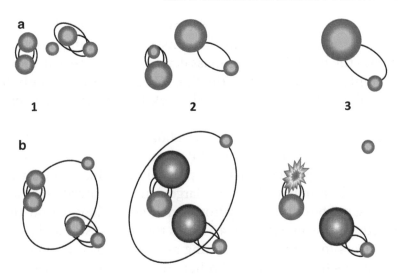

Fig. 6.2 The effect of stellar density on the evolution of compact stellar groups. In (**a**), stars are grouped closely and the frequency of collisions is high, yielding a few ultra-massive stars (**3**). In (**b**) the stars are further apart, and collisions between stars are rare. Individual massive stars have a chance to evolve away from the main sequence (**2** and **3**) before collisions occur. Extensive mass loss in these stars also encourages further separation between stars. According to Simon Portegies-Zwart the clusters MGG 11 and MGG 9 in M82 may be analogous to models (**a**) and (**b**), respectively

in MGG 11. The stellar density in MGG 9 is sufficiently low that stars have enough time to dither in the outer parts of the cluster and avoid the collisions that lead to the formation of freakishly massive objects (Fig. 6.2). Stellar evolution takes them away from the main sequence before collisions become likely. These stars then evolve into neutron stars or stellar mass black holes through standard stellar evolutionary processes.

In 2011 it was reported by Amy Reines and colleagues that the nearby dwarf irregular galaxy Henize 2–10 also appeared to host an actively accreting black hole. The analogy with M82 was clear. Henize 2–10 is also undergoing a very active starburst, and the galaxy contains many massive clusters of stars. However, one such cluster, positioned near the galactic center, hosted an object that produced an unusual combination of hard X-ray and strong radio wave emission. The radio luminosity is 7.4×10^{35} ergs^{-1} at 4.9 GHz, and the X-ray luminosity is 2.7×10^{39} ergs^{-1} in the 2–10 keV. Although not stunning individually, it is their combination that is suggestive. Given these values the mass of the black hole can be calculated, assuming that all the emission is coming from accretion onto the black hole. The mass derived from these calculations is one million times that of the Sun. Henize 2–10 may only be a dwarf galaxy, but it has a central black hole to rival that of the Milky Way.

In 2007 when the light of SN 2006gy was still fading, Portegies-Zwart and colleague, Edward P. J. van den Heuvel (University of Amsterdam), proposed that the progenitor of this supernova was the result of a merger between larger, more evolved stars and a still massive but smaller hydrogen-rich star. They reason that a star with a mass in excess of 40 times that of the Sun will shed its outer layers long before detonating as a supernova. Leaving aside observations by Gal-Yam and others that massive stars can die as LBVs (Chap. 3), Portegies-Zwart and van der Heuval proposed that a collision could give rise to a massive, highly evolved but still hydrogen-rich star.

The essence of their model for SN 2006gy is that approximately 100,000 years ago a collision occurred between a 20 solar mass star and another behemoth containing 100 solar masses or more material. Indeed, this second object possibly had almost 1,000 solar masses, as the result of frequent stellar collisions in the first few million years of its life.

This final event was, in effect, the work of a makeup artist. The collision provided a thick veneer of foundation that covered the signs of decrepitude in the more massive progenitor. This highly evolved, but seemingly more youthful, object, then blew itself apart as SN 2006gy. The time lag between merger and detonation allowed most of the 20 solar mass object to be dispersed through stellar winds. These, in turn, provide the thick shell of hydrogen-rich gas around the supernova that instigated the magnificent output of the explosion when the shockwave hit it.

Although more speculative than the pulsational pair instability model (Chaps. 3 and 8), the theory is utterly reasonable on the grounds that collisions occur within star clusters and can be frequent. Moreover, the authors propose a test of the model – the emergence of a 5-million-year-old dense, massive star cluster in the location of SN 2006gy. Such a cluster should be apparent once the supernova fades. This should contain around 10,000–100,000 solar masses, of which much would be found in very massive stars. Observations of sufficient resolution could be made to allow testing of the merger model, assuming that the density of the cluster can be ascertained. This is a big assumption. The presence of a star cluster alone is insufficient, as most massive stars are born in groups. This is a case where time will tell.

Conclusions

Although relatively rare, in the right circumstances stellar collisions can yield exotic objects that are capable of undergoing core collapse. In some instances the outcome of a collision may yield an unstable object called a Thorne-Zytkov object (TZO), which should collapse promptly to form a black hole – and perhaps an accompanying GRB. In other cases stellar collisions within dense star clusters may produce such a fiendishly massive object that it collapses to form an intermediate mass black hole.

Much of the work is theoretical, requiring many days of intense computer time. Observations of collisions are rare (Chap. 14), and much of the evidence is indirect. Given the scarcity of massive stars, the chances of observing the collision between these stars will be less likely still. Moreover, resolving the collision and its product will be made difficult by the sheer brilliance of the stars and their tight association in the cluster.

Astronomers still aren't sure what R 136a is. It may be one 260 solar mass star or a tight binary (or multiple) star system. The combined visible output is simply too great to be sure what is generating it. We may have to wait for a lower probability event, such as a collision in a binary system or a merger in a more open star system (Chap. 14). These are inherently less likely than the equivalent collision in a dense cluster, but they will be a lot easier to observe and analyze.

Hopefully, the increases in automation will allow the rarest of rarities to be observed – a stellar collision within a cluster of massive stars. What will be needed is a large, optical 'scope equipped with adaptive optics and a healthy heap of luck. Either way the impact of one star upon a neighbor is likely to be dramatic and worthy of further investigation.

7. Electron-Capture Supernovae

Introduction

What have some stars got in common with gamblers? Admittedly the answer may not seem obvious at first. However, the clue to solving this riddle lies with the fate of a class of transitional stars. These celestial bodies demarcate an evolutionary fence, separating the core-collapse supernovae and the formation of planetary nebulae. They present huge computational headaches for astrophysicists. Their fate is determined by two competing forces: core growth and mass loss from the bloated envelope of the star. Both forces are acting at very similar rates in these finely balanced stars, and the outcome depends on which of the two forces wins out.

These stars are like pencils balanced on their tips and consequently are challenging to model. However, if these stars can explode as supernovae, it may explain some unusually weak supernovae, as well as chemically peculiar Type II events such as that which formed the Crab Nebula in A. D. 1054.

To illustrate the unique and very controversial fate of these transitional stars, we introduce two peculiar explosions, SN 2008S and SN 2009md. The nature of SN 2008S in particular emphasizes the degree of controversy.

Supernova or Imposter?

SN 2008S was a peculiar beast. Caught brightening on February 1, 2008, this hydrogen-rich "supernova" never attained a luminosity comparable with most Type II-P events. Indeed its peak absolute magnitude was −13.9, placing it clearly at the tail end of observed supernovae, but well within the range of so-called "supernova imposters," "Type V supernovae" or η-Carinae analogs. In particular SN 2008S was typical of so-called "faint Type IIn supernovae," displaying the characteristic narrow emission lines in its spectrum, but unusually dim by the standards of core-collapse events. The question then arose, are these real and very terminal explosions, or are they something less violent that can be understood in terms of the evolution of a massive star?

Underlining this controversy, Nathan Smith interpreted SN 2008S as a supernova imposter – a brief, violent, but non-terminal explosion of a massive star – while José Prieto ascribed the event to an "electron-capture supernova." In order to flesh out these possibilities we need to understand the complex fate of stars in the range of 7–9 solar mass range, while comparing their fate to the brief, violent but non-terminal explosions encountered in Chap. 3. Once we've delved into this developing mystery, we will return to the fate of SN 2008S and consider which model is most reasonable for this diminutive supernova. As we travel along this path we will encounter those fateful gamblers who've inadvertently contributed to the science of supernovae.

The Troubling Fates of Intermediate Mass Stars

We'll begin this story by considering how stars 7–9 times the mass of the Sun evolve and die. This is a surprisingly complicated tale and one which, despite decades of work, is still wracked by uncertainty. The principle problem in modeling these stars arises because they are so finely balanced between one fate and another. Getting the figures even slightly wrong will make for a dramatic difference in the outcome of any evolutionary models. As we delve deeper into the fates of these stars, these problems will become clear.

An intermediate mass star generally follows a similar evolutionary path as the Sun. However, such stars are considerably hotter, brighter, bluer and live comparatively short lives that are measured in at most a couple of tens of millions of years. Like the Sun, the cessation of core hydrogen burning is followed by the ascent of the red giant branch, core helium burning and a second, or asymptotic, giant branch phase that culminates in the loss of the hydrogen-rich envelope and the death of the stellar core as a white dwarf.

Near the top of this intermediate mass range (7–9 solar masses) stars are sufficiently weighty to generate the core temperatures needed to burn the carbon produced previously by helium fusion. Models suggest that a proportion of these stars can, in principle, undergo core collapse, triggering a supernova. In turn, the profuse loss of mass that these stars undergo as they approach death has generated interest in the astronomical community in those looking to explain weaker Type II events.

Before we can consider the merits of the case for those stars producing certain Type II explosions, we need to understand how they evolve away from the main sequence and how this in turn can lead to supernovae.

Intermediate mass stars, in the range of 7–9 solar masses, are able to burn carbon in their cores. Carbon ignition occurs under nearly degenerate conditions as the star enters the asymptotic giant branch (AGB phase). The term "degeneracy" refers to the properties of the electrons present in the stellar core. In degenerate materials, the electrons line up in respective energy levels that allow the material to resist further collapse. It is this process that halts stellar collapse in white dwarf stars and ultimately brings about their slow demise as terminally cooling objects. Carbon fusion produces neon and magnesium, while the oxygen remains untouched.

Degeneracy means that when the core ignites carbon under these conditions it can result in explosive instability, akin to the helium flash in stars such as the Sun. Why? When a material is degenerate it resists the sorts of usual and quite tolerant behaviors of most materials. Most importantly, a non-degenerate material, whether a solid, liquid or gas, will expand merrily when heated. Degenerate materials adopt a much more stubborn outlook and thoroughly resist expanding unless given a very violent "kick up the proverbial." Therefore, should the degenerate material dare to undergo nuclear fusion, the energy generated is initially trapped, causing an upward spiraling of temperatures and nuclear activity. Consequently, the ignition of carbon is an aggressive affair that both shakes and stirs the stellar interior, causing significant alteration in the star as a whole.

However, even as the core undergoes its own spasms the outer layers of the star are wracked by powerful stellar winds. These are whipping away more than a hundred thousandth of the star's mass every year. That may not sound much, but this rate of mass loss could reduce a 10 solar mass star, with the bulk of Spica, to a few wispy specks in a few hundred thousand years – a blink in the cosmic eye.

Ignition of carbon confers a brief hiatus because the rearrangement of the stellar interior causes a contraction and heating of the stellar envelope. On the HR diagram the star executes a so-called blue loop, briefly leaving the red giant branch and retreating left to higher temperatures. With a smaller diameter, the stellar core can maintain a more assertive grip on events happening at the stellar surface, and the rate of mass loss briefly plummets. Carbon fusion produces neon and magnesium, and the inner core becomes enriched in these elements plus the oxygen left over from helium fusion. It turns out these elements can have a profound effect on the subsequent fate of the star if the core grows sufficiently large.

After only a few hundred years the stellar core exhausts carbon and the whole sorry business of mass loss repeats. The core contracts and heats up, driving the re-expansion of the star. The renewed expansion re-instigates a final, vigorous period of mass loss from the exterior, and, once again, the star loses control of its fate.

However, if there is sufficient mass in the hydrogen and helium layers the core can become hot and massive enough to ignite the neon produced by carbon fusion and push on to the fate bestowed by iron disintegration – core collapse. In contrast, if the star has lost too much mass before the neon ignites, stellar evolution is terminated by those strong winds that rip away the remainder of the envelope. The question is, can a stellar minority teeter on the brink between these fates, like a pencil balanced on its tip? The star quietly builds up its carbon-depleted core to the point at which it collapses due to a process called electron-capture.

Limiting Factors

Before we investigate the mysterious process of electron-capture it's worth examining what evolutionary force limits the growth of the neon core, preventing the headlong rush to iron-core collapse.

As previously mentioned, the precise fate of stars born with 9–9.25 solar masses has interested astronomers for decades. Modeling the lives of these stars is complex primarily because they really do teeter on a knife-edge. Lose too much mass and they become oxygen-neon-magnesium (or ONeMg) white dwarfs, but lose too little and they pursue the well-trodden path of more massive stars, ultimately exploding as Type II supernovae.

The problem is twofold. On the one hand, the force that whittles away at the envelope from the outside – mass loss – is as poorly understood as the process that whittles away at the envelope from the inside – nuclear reactions. Unless you know exactly which rates are which you can't balance mass loss accurately against core growth. Clearly, without this, and to put it bluntly, you're stuck.

The initial key to understanding this fate was modeling work done by Icko Iben Junior and colleagues (Enrique Garciá-Berro, Claudio Ritossa) in the late 1990s. Iben built on earlier work on these transitional mass stars carried out by Ken'ichi Nomoto. In 1984 Ken'ichi showed that stars in the mass range of 9–12 solar masses would develop a degenerate oxygen-neon-magnesium core following carbon burning and end their lives as oxygen-neon-magnesium-rich white dwarfs. Iben and colleagues then extended the analysis, examining a succession of models of stars with 8–11.5 solar masses. At the lower end of the mass range, carbon burning produced the oxygen-neon core, but it could never grow big enough to ignite neon and move forward to iron-core collapse. However, in Iben and colleagues' models, stars at the top end of the range (11–11.5 solar masses) were able to grow the mass of the helium-depleted core to the point at which it could collapse through the exotic electron-capture mechanism.

Subsequent work by Arend Poelarends and Alex Heger has refined the calculations used in the earlier work and shown that the critical mass range for SAGB evolution is likely to extend from 7–9.25 solar masses, somewhat lower than Iben and colleagues initially proposed. The obvious difference in the mass range (and the one that is used here) reflects better understanding of the rates of nuclear reactions, mass loss and the inclusion of stellar rotation in the models analyzed.

In the 1990s Iben used non-rotating stars to cut down on the amount of computations needed to run the models. However, clearly, real stars rotate, and when rotation is taken into account the core mass can grow larger because of mixing of hydrogen from the envelope into the stellar core while the star is on the main sequence. Furthermore, improvements in the understanding of processes such as convection, mass loss and what are called dredge-outs (described in detail below) mean that more realistic models have now been derived.

Post Main Sequence Evolution

In Heger's and Poelrand's work the fate of an intermediate mass star in the range of 7–9.25 solar masses rests with dredge-up (or dredge-out, as it is sometimes referred to). Dredge-up is a phenomena linked to deep convection in the hydrogen-rich envelopes of massive stars. Low and intermediate mass stars develop deep convection in their hydrogen-rich envelopes as they evolve off the main sequence, as well as at later stages. In stars such as the Sun, this will be a relatively modest affair. However, in the more massive stars this dredging effect can extend down to the helium core. Stars with more than four times the mass of the Sun experience three episodes of dredge-up that fundamentally alter the chemical composition of the stellar envelope and ultimately the evolution of the star.

The first episode of dredge-up occurs as the star evolves away from the main sequence and ascends the red giant branch for the first time. This first phase commences as the inert helium core contracts and heats up. Hydrogen ignites in a shell around the core and continually adds mass to the core. All the while the outer layers are expanding and cooling, deepening the extent of convection from the outside inward. As the star swells, convection digs deeper and deeper down toward the hydrogen-burning shell surrounding the core of the star.

Eventually the products of hydrogen fusion begin to appear at the stellar surface as the two layers meet. At this stage dredge-up is manifest in these stars as a steady lowering of hydrogen abundance at the star's surface, enrichment in the abundance of helium produced by earlier rounds of nuclear fusion, and most importantly the appearance of the products of carbon-nitrogen

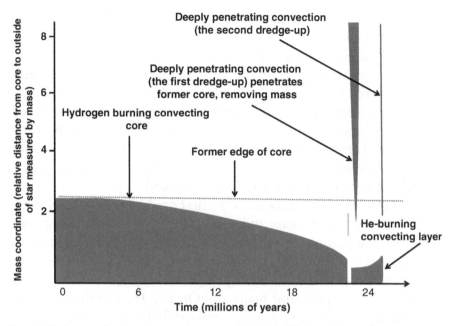

FIG. 7.1 Areas of core convection are shown in blue and envelope convection in *red*. The most significant periods occur at 23 and 25 million years after birth. Here convection currents reach down to the core and dredge material upward (Adapted from "The evolution of an intermediate mass star" by Ritossa, Garcia and Iben, *Astrophysical Journal*, 1996)

cycle nuclear reactions. Consequently, during the first dredge-up the abundances of carbon and nitrogen isotopes in particular are altered. Most significantly the abundance of nitrogen isotopes (principally nitrogen-14) is enhanced as these isotopes are brought up from the hydrogen-burning shell at the edge of the core. Meanwhile some carbon isotopes (carbon-12 in particular) become rarer. The exchange in abundances relates to the role carbon plays in hydrogen fusion (Fig. 7.1).

In the carbon-nitrogen cycle, carbon serves as a catalyst in the creation of helium from hydrogen. However, in the various reactions that create helium, some are relatively slow compared to others. The conversion of nitrogen-14 to oxygen-15 is one such step. Consequently, as the amount of hydrogen runs down in the core, more and more carbon-12 ends up locked up as nitrogen-14 rather than completing the round of fusion that would regenerate the carbon-12 and liberate helium.

Although interesting from a chemical point of view, changes in the star's chemistry are a sideshow. What really matters is the effect of the first dredge-up on the growth of the helium core. Should the dredge-up limit the core's growth, its impact would pose problems for later stellar evolution. Remember that the fate of the star is linked to its available fuel. If the hydrogen-depleted core stays small, nothing more exciting than helium burning will occur. Fortunately, the impact of the first dredge-up on core growth is minimal, and the core easily reaches the mass required for helium ignition (roughly 0.5 solar masses).

For intermediate mass stars the first dredge-up only lasts a few tens of thousands of years and is insufficient to affect core mass. However, subsequent intrusions of the convecting outer layer of the star have a more drastic influence in the star's subsequent evolution. Once the core has ignited helium, the star readjusts its structure, shrinks and heats up. On the Hertzsprung-Russell diagram it executes a blue loop, moving left toward higher temperatures but lower luminosities. This phase is sustained for a few million years while the star burns helium. With the star smaller and denser, gravity is able to restrict mass loss from the stellar surface, and the star takes on a brief semblance of stability.

Despite a minimal impact of the first dredge-up on stellar evolution stars with 4 or more solar masses experience a second dredge-up once the core is depleted in helium. It is this incursion that is particularly important for the more massive stars in this range.

Once again, as core helium fusion ceases the core contracts and heats up, while the envelope of the star expands, cools and the depth of convection increases once more. A star evolving through this second, or asymptotic, giant branch phase encounters second dredge-up early in this phase. The star moves back toward the red giant branch, steadily brightening while the rate of mass loss accelerates once more. As convection deepens it causes the base of this outer convecting layer to bite down into the hydrogen-free core. As it does so, once again some of the remaining inert helium is removed and transported into the expanding stellar envelope. This second dredge-up lowers the mass of the stellar core because it removes helium faster than it is replenished by the rather worn-out hydrogen-burning shell. Indeed some models show that hydrogen fusion is initially switched off during this phase. This allows easy access of the envelope's convection cells to the leftovers of the helium core. Loss of mass from the helium and carbon-oxygen core immediately restricts the time the star can spend subsequently evolving.

If the helium-free core mass stays over 1–1.2 solar masses carbon can ignite and carry the star forward through a third stage of nuclear burning. For the most massive stars in this range (7–9 solar masses) the second dredge-up can limit core mass so much that although carbon is burned the products of these fusion reactions, neon and magnesium, are not. In this case stellar evolution terminates once the carbon is used up and stellar winds remove the outer layers of the star. Poelrand's and Heger's work shows that the second dredge-up can reduce the mass of the hydrogen-free core by enough to prevent the core from reaching the Chandrasekhar limit or consequently igniting the neon produced by carbon fusion.

Why is this important? For the more massive stars, the second dredge-up prevents neon ignition because it limits how large the carbon core can grow. Without neon fusion no further rounds of nuclear fusion take place, and the formation of a massive iron core is blocked. This implicitly prevents the star from entering the route to "conventional" core collapse and Type II supernova. For those lower mass stars in the 7–9 solar mass range, the influence of the second dredge-up is sufficient to prevent core growth to the Chandrasekhar limit. Subrahem Chandrasekhar showed that cores of a certain mass could not support themselves with electron degeneracy and would collapse if the mass exceeded this limit. For a core composed of oxygen, neon and magnesium this is around 1.382 solar masses. It is somewhat higher for cores composed of iron. Consequently, if the second dredge-up is too severe, the growth of the core is blocked and neon will never ignite, nor will the core reach the mass necessary to undergo collapse.

The Cassino da Urca

What, then, is the link between gambling and supernovae? The answer to this question has its origin in the 1950s. George Gamow and Mario Schoenberg were visiting a casino in Rio de Janeiro named the Cassino da Urca. They observed gamblers entering and departing the casino with a deplorable repetition, and it is said Schoenberg quipped to Gamow that "The energy disappears in the nucleus of the supernova as quickly as the money disappeared at that roulette table."

So what exactly is the link between gambling and supernovae? The answer lies in the ephemeral neutrino and the propensity of some nuclei to undergo exotic nuclear reactions when they are compressed to high density. There are two connected Urca processes, as they are known, that are linked to supernovae. The first affects cooling in neutron stars – the end products of most core-collapse supernovae, while the second connects to cooling processes affecting white dwarf stars. As we can think of the leftover oxygen-neon-magnesium degenerate cores as white dwarfs, the role of these Urca processes also mediates the fate of 9–9.25 solar mass stars.

An Urca process involves the absorption of an electron by a nucleus deep in the core of a massive star. Absorption of an electron results in the conversion of a proton to a neutron and the release of energy in the will o' the wisp form of a neutrino. This ghostly particle merrily zips out of the core, removing energy. The process of electron-capture also removes electrons from the inner portions of the core – those most in need of their support against core collapse. This encourages the collapse of the core and an increase in density, which in turn accelerates Urca processes. Therefore, Urca processes are a nasty double-edged sword. Both blades cut deep into the star, triggering its death.

In order for the Urca processes to happen efficiently, the material has to be degenerate so that the position of electrons around the nuclei predisposes them to absorption onto their parent nucleus. Only a few isotopes of a few elements will suffer this fate – and notably it is some of the products of carbon fusion that have this predisposition. Therefore, at this evolutionary stage, the core is full of these elements and their isotopes, and Urca processes are a natural and unavoidable consequence (Figs. 7.2 and 7.3).

A true Urca process has two stages unified by the process of convection (see Figs. 7.2). The electron absorption step occurs in the deepest reaches of the core, and it is followed by the removal of the product nucleus by convection to cooler and less dense

> **Urca Processes**
>
> Urca processes are paired electron-capture and release reactions that result in energy losses in very dense, degenerate materials. Urca reactions are important mechanisms that allow degenerate materials to cool efficiently and therefore have a special place in the final stages of stellar evolution where degeneracy terminates the lives of low, intermediate and the majority of massive stars.
>
> Urca reactions occur in pairs that complete a cycle and regenerate the original isotope. In the process energy is lost by the emission of neutrinos and antineutrinos.
>
> Step 1: The capture of an electron by the nucleus of the atom and a proton is converted into a neutron. Electron capture is accompanied by neutrino emission.
>
> Step 2: Convection carries the newly minted neutron-enriched nucleus upwards to a less dense environment, where it undergoes beta decay. An electron and an antineutrino are released as the neutron decays to form a proton.
>
> If conditions are suitable the process of convection can carry the isotopes undergoing Urca reactions repeatedly into and out of the densest parts of the stellar core. Loss of energy in the form of neutrinos and antineutrinos removes energy from the core, and the absorption of electrons in the inner part of the core removes some of the degeneracy pressure needed to keep the core supported against gravity.
>
> In the early stages of neon core cooling aluminum-27 absorbs an electron, forming magnesium-27; magnesium-27 then decays, releasing a beta particle (essentially an electron from the nucleus). In the core of a star this reaction has limited importance, as aluminum-27 is relatively rare.
>
> As core collapse continues and densities increase neon and magnesium isotopes serve as components of Urca chains. Since these are abundant – the products of carbon fusion – electron capture onto these is catastrophic, robbing the core of support and ultimately triggering core collapse. Of particular importance is the absorption of an electron by neon-20. Neon-20 converts to fluorine-20, which then undergoes decay, releasing another beta particle and antineutrino while regenerating the original neon-20 isotope. The half-life for these isotopes is in the seconds to minutes range, and although this sounds insubstantial it is sufficient to allow core collapse once the process becomes efficient at suitable densities.

FIG. 7.2 A run-through of the processes that ultimately rob the oxygen-neon core of the star of its internal support. These Urca processes are complex and are only summarized here, but they may be the driving force for 4 % of core-collapse supernovae

regions within the core. The new, neutron-enriched nucleus is thus transported outward by convection to more salubrious surroundings, where it then undergoes beta decay. In simplest terms, the electron that was absorbed earlier is now released, regenerating the earlier element and transporting the electron to the outside of the stellar core. As the material convects, the isotopes needed for the Urca process to occur are continually regenerated, swinging in and out of the deepest regions of the core. The net result is a steady reduction in the number of electrons present in the heart of the core. Unfortunately, as the core is supported by electron degeneracy pressure, removing electrons is a bad move. This, coupled to energy losses by the release of neutrinos, is what instigates core collapse.

In Icko Iben's work the 11 solar mass star (broadly equivalent to the 9.25 solar mass object in Heger's and Poelarends' work) has

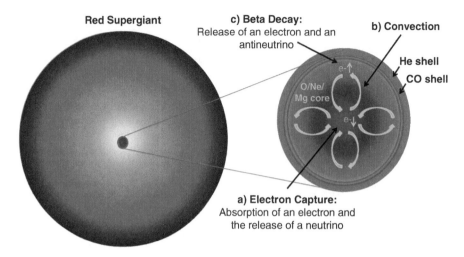

FIG. 7.3 A gambler's death. The core of the 9.25 solar mass star plays its final hand at the Casino da Urca. Nuclear reactions in the heart of the neon core (**a**) remove electrons, generating neutron-rich isotopes of various elements. These reactions sap energy from the core through the release of neutrinos. Convection (**b**) drags these neutron-enriched isotopes up to the top of the core where they decay, releasing an electron once more and an antineutrino (**c**). Over time, these reactions make the core denser until it finally collapses to form a neutron star. The stellar core is not drawn to scale

a very brief flirtation with fusing neon and also oxygen. However, energy losses by neutrinos soon cool the shrinking core below the temperatures necessary to maintain this.

In the degenerate carbon-depleted core of the moribund 11 solar mass star of Iben and colleagues, Urca processes steadily convert one suitable isotope to another and, as the mass of the neon core increases, more and more isotopes become involved in these wasteful reactions. Urca processes also reduce the temperature of the core from a toasty 700 million to 175 million Kelvin, as they transfer energy outward. All the while electrons are flowing inward, absorbing steadily isotopes of aluminum, then magnesium and sodium, and finally isotopes of neon. Each reaction requires a greater density than the preceding one, and each one becomes more and more frequent as the abundance of these isotopes increases.

The transfer of heat from the central to outer portions of the core drives an outward-moving convective shell through the helium-depleted core. After 30 years the core is severely depleted in the isotope aluminum-27, as well as many others, and is ready for final collapse.

At the end of this phase, carbon is severely depleted through nuclear fusion in the core of the 11 M star (much less than 0.001 % by mass). The depleted core also contains an interesting assortment of nuclei created through Urca reactions that are normally unstable in terrestrial environments (unusual isotopes of elements with mass numbers 23, 25 and 27). Iben and co-workers' calculations showed that the final core has 1.263 M of oxygen and neon, surrounded by a thin shell consisting of 0.0082 of carbon and oxygen; this is overlain by a wistful helium shell of 0.00065 M (considerably less than the mass of Jupiter).

These translate to a nearly bare core overlain by steadily thinning shell of hydrogen in the 9 solar mass models of Heger and Poelarend through to a more bulky hydrogen envelope in the 9.25 solar mass star in their models. It must be stressed that the original models of Iben and co-workers were non-rotating and therefore less realistic than the later models by Alexander Heger and colleagues. Although the original work by Iben is breathtaking in its completeness, and in terms of the insights into the remarkably different fates these stars have, they depend on relatively narrow windows in their mass.

After a few decades the hydrogen-burning shell has increased the mass of the hydrogen-depleted core to 1.382 solar masses. At this point the density is so great that neon-20 takes part in Urca reactions. As neon-20 is an abundant component of the core, its loss obliterates electron degeneracy within the core, causing it to collapse. The fatally destabilized core promptly implodes at 70,000 km/s. A fraction of a second later a 1.2 solar mass neutron star has formed, and the envelope and outer core are ejected in a weak Type II supernova.

The Type II supernova that accompanies core collapse has been called "silent" by Iben and colleagues. The amount of hydrogen ejected in the explosion is lower than normally seen in the explosions of more massive stars. This is due to preceding phases of enhanced mass loss that accompanied core helium and then carbon exhaustion. Consequently, the ejecta are expected to be richer in helium than standard Type II events, and the duration of any plateau phase in the light curve may be similarly limited.

What else might help astronomers distinguish these electron capture supernovae from other Type II-P events? In Heger's and Poelarend's 9 solar mass models the hydrogen shell is expected to have been largely stripped away by the time the core collapses. The supernovae that result might be apparent as Type II-L events (those showing a faster, linear decline in light compared to Type II-P), or they may appear as so-called Type IIb events. These are explosions that initially show hydrogen but soon lose this to reveal helium and other elements from deeper portions of the shredded star. By contrast, a few of these events may appear as Type IIn explosions where the stellar winds have stripped material away from the dying star at a rate of ten-thousandth of a solar mass per year, and this swathe of matter impacted by the blast wave from the subsequent supernova.

Calculations by numerous groups also suggest these explosions are relatively weak, in the region of 4×10^{49} erg, or several times weaker than an average Type II-P explosion (10^{51} erg). The amount of radioactive nickel-56 is also depleted relative to other core-collapse supernovae. Typical amounts produced in models lie in the region of 0.003 M or less than 1/10th that seen in typical core-collapse supernovae. Furthermore the elements synthesized in these explosions are expected to differ from other Type II explosions. Work by some researchers indicates that these supernovae are deficient in so called r-process elements (those produced by the rapid addition of neutrons) but mildly enriched in elements such as zirconium, zinc, arsenic and rubidium. These elements are produced in a different series of reactions called the p-process, where protons are added quickly to seed nuclei rather than neutrons.

A final and interesting observation relates to the final product of these explosions, the neutron star. There are strong hints that the masses of neutron stars cluster around two peaks: 1.25 solar masses and 1.44 solar masses. This double-peaked, or bimodal, distribution of mass hints at two distinct origins for these stars. One suggestion, made by Philipp Podsiadlowski, is that the lower peak represents neutron stars formed by the electron-capture route, while the second peak represents the iron-core collapse route.

Moreover, Christian Knigge and Malcolm Coe (University of Southampton, UK) have analyzed a large population of X-ray binaries, and their work suggests that not all neutron stars are born equally. Knigge and Coe utilized the large population of these binaries found in the Small Magellanic Cloud, as well as smaller populations in the Milky Way and the Large Magellanic Cloud. BeX binaries consist of a fairly massive and fast-spinning B star in an elliptical orbit with a less massive neutron star. As the neutron star makes its closest approach to the B star it slices through a disc of material surrounding the stellar equator. A pulse of X-rays is then generated as the neutron star sweeps up some of the gas and heats up. By measuring the pulsation in the emission of X-rays, the team was able to discern two distinct populations of neutron stars. One population had an orbital period peaking around 10 s, while another peaked at around 5 min. This suggested two different formation routes.

The presumption, which awaits further confirmation, is that the long-period binaries were formed when one star generated a massive iron core that then collapsed. The kick velocity from the explosion was sufficient to drive the nascent neutron star into a more distant, elliptical orbit. The second, shorter period population was derived from electron-capture supernovae. These explosions are predicted to impart a far smaller birth kick to the neutron star, leaving it in a tighter orbit around its companion. More work is needed, but despite a lack of clear observational evidence for electron-capture events, their aftermath appears in the cosmological record.

Finally, the models of Alex Heger and others suggest that 4 % of core-collapse supernovae could be electron-capture events. There is notable agreement between measurements of element and isotope abundance and those predicted for electron-capture explosions, given the expected rate of their occurrence. This may be coincidence, but it certainly adds another piece of circumstantial evidence supporting the formation of supernovae through this mechanism.

As astronomers scramble to definitively identify these a number of interesting supernovae have come to light. Some suggestions for these electron-capture events include the weak Type II-P explosions SN 1994N, 1997D, 1999br, 1999eu, 2001dc and 2005cs. The true cause of these weak supernovae remains to be determined. Whether these supernovae are generated by electron-capture, fall-back onto a central black hole or through the accretion-induced collapse of a white dwarf remains to be determined. However, at present there are a growing number of supernovae that appear to fit the bill for electron-capture events. Further analysis is warranted to confirm or refute whether these claims are justified.

Many Roads Lead to Rome

Massive single stars in the range discussed may constitute the majority of stars that encounter electron-capture, but they are not the only ones. Another class of progenitors are the massive ONeMg white dwarf stars left over after carbon burning but were insufficiently massive to collapse in their own right.

Given the right circumstances, a pairing of the ONe white dwarf with another star – either a main sequence star or red giant – can allow the white dwarf to accumulate matter from its companion through an accretion disc. If the white dwarf star gains sufficient mass, it will also undergo electron-capture and then collapse. These events, called accretion induced collapse (A. I. C. for short), may contribute a pool of core-collapse supernovae that are observable. The properties of these events may be very different in terms of light curve and nucleosynthesis, depending on the type of matter that is accreted and at what rate. For example some white dwarf stars may accrete hydrogen from a companion at a high rate, and their explosion may bear some resemblance to Type IIn events as the blast wave punches through the accretion disc surrounding the dwarf. Other white dwarfs might accrete helium-rich material from a much evolved giant companion or from a lower mass white dwarf. The products of helium fusion may appear in the supernova as calcium-rich debris. The bottom line is that there are many possible routes to this kind of supernova, depending on the specific system in which the white dwarf is found and consequently many outcomes in terms of spectra and light curve. These are returned to in Chap. 13.

To this pool of relatively massive stars, we can add the accretion-induced collapse (A. I. C.) of oxygen-neon white dwarf stars. The outcome of stellar evolution for all those other 7–9 solar mass stars are white dwarfs enriched in oxygen, neon and magnesium. If these corpses find themselves in a tight enough binary with a main sequence red giant star they can accrete matter. In the right circumstances, and with enough mass transferred to prevent erosion in thermonuclear novae blasts, the white dwarf will steadily grow in mass. Once it reaches the canonical limit set by Subrahem Chandrasekhar (1.382 solar masses) the core implodes. Without carbon to burn and trigger a Type Ia supernova, the core collapses to become a neutron star, and the outer layers are ejected in a Type II supernova – assuming it is hydrogen that has been accreted. If, instead, helium is accreted from the companion, the supernova might appear as a Type Ib event.

Did the Progenitor of SN 2008S Spend Too Much Time at the Roulette Table?

With the fate of intermediate mass stars in mind we return to the faint SN 2008S – an explosion bearing the hallmarks of a Type IIn event. However, does the evidence point towards an electron-capture supernova, or was it something else? This is what we know: SN 2008S had a peak luminosity of less than 10^{48} ergs (10^{41} J); it ejected less than 0.2 solar masses; the progenitor, identified by using the Spitzer Infrared Telescope, was heavily obscured by dust; and finally, the spectrum revealed narrow emission lines, indicating that the eruption bore a kinship with Type IIn supernovae.

José Prieto (Princeton University) and colleagues interpreted this explosion as an electron-capture supernova, occurring within a SAGB star that was heavily obscured by the dust of its own creation. This material lay in an extended shell stretching from 90 to 450 A. U. from the central star. Subsequent work by Maria-Teresa Botticella (Queen's University, Belfast), adopted the same model for this supernova. They pointed to an inferred infrared luminosity of 35,000 solar units for the progenitor star, based on earlier infrared Spitzer observations of the progenitor's location. By comparison with stars on the HR diagram, this output put the star at around 10 solar masses. The low luminosity and observed sluggish blast wave (moving at less than 1,000 km/s) were a consequence of the progenitor's low mass and the electron-capture route to detonation. Certainly, the luminosity is compatible with a star of this mass, assuming all of the radiated energy of the star is escaping at infrared wavelengths. However, is there an alternative explanation that can explain the other features of the supernova?

Given the same information Nathan Smith and co-workers (University of Arizona) saw SN 2008S in a rather different light. SN 2008S wasn't a true supernova but was an example of a supernova imposter, a non-terminal explosion occurring on a much more massive (20–30 solar mass) star. This was something similar to the eruptions of LBV stars (Chap. 3). The thickly blanketing dust was interpreted as the remnants of a recently ejected red supergiant shell as the star executed a blue loop on the HR diagram. As we have seen, these loops to higher temperatures may be caused by the ignition of core helium or carbon, depending on the mass of the star, followed by the readjustment of the internal structure of the star. Indeed, in the final chapter we examine a blue transient – a star that appears to have suddenly brightened and ejected a modest amount of its mass in such a scenario. Therefore, luminous but non-fatal outbursts are possible in stars that are merely going through temporary but dramatic changes in their evolution.

In the imposter scenario the observed explosion occurred as SN 2008S progenitor shrank and heated dramatically, driving a faster wind through the dissipating shell of matter from its previous red supergiant phase. The pulse of energy, observed as the imposter, was produced in the collision of this faster blue (or yellow) supergiant wind with the denser, slower-moving and older red supergiant wind. If so, the star is still intact, and this was not a true supernova. The star should then emerge from its cocoon in a few years, reborn as a yellow or blue supergiant.

What evidence is available to allow us to discriminate between the two scenarios? The slow expansion speed is certainly mediocre to say the least for any supernova and has much more in common with the sorts of violent burps and bangs driven by massive luminous blue variables (Chap. 3). Similarly, the relatively small ejected mass would suggest a violent, but non-terminal, event in a massive star. Finally, the overall luminosity of the explosion (less than 10^{48} erg) is mediocre even by the expected standards of an electron-capture supernova, coming in at roughly a tenth of what might have been expected from this route. Conversely, the estimated infrared luminosity of the progenitor seems too low for a star of 20–30 solar masses and would appear to favor the electron-capture scenario.

However, the estimation of total progenitor luminosity is based solely on Spitzer images, and the lack of evidence for a bright object at visible wavelengths. It isn't clear whether all the radiated energy of the parent star is leaving the dusty shell that surrounds the star. If it isn't then measurements of stellar mass based on infrared data alone will underestimate total luminosity and hence the mass of the parent star.

There is one final possibility that would be consistent with both a high progenitor mass and a low luminosity of the explosion: the death of a massive star and the concomitant formation of a black hole. In this scenario the blast wave is slow because little energy was imparted into the blast wave, and the majority of the ejected material then fell back onto the young black hole or neutron star before its final collapse. Again, this scenario might be distinguishable from the other two by the presence of an anomalous X-ray output associated with accretion onto the central black hole or through the formation of a jet. As neither has been observed, it implies that this scenario is improbable. However, it is worth considering each possibility in turn. Moreover, it was gratifying to note a clear collaboration between the authors of the two conflicting papers analyzing this explosion. This very healthy state of affairs should be encouraged.

Returning to SN 2008S, the proof of the pudding may yet be in the tasting. If SN2008S is a bona fide supernova the blast wave should eventually push aside the dust shell surrounding the star, revealing a hot X-ray emitting core. If, however, the event was akin to an outburst of the yellow hypergiant rho Cassiopeia or luminous blue variable, the parent star should emerge once more, somewhat bluer, somewhat bashed but still gloriously alive. Time will tell.

SN 2009md: A Faint Type IIP Supernova with a Troubling Origin

Supernova SN 2009md was certainly just that – a supernova. Unlike SN 2008S, which bears more than a passing resemblance to an imposter, this supernova clearly belonged to a small but unique class of faint core-collapse but hydrogen-rich events. Most importantly, not only was the supernovae caught, but the progenitor of this explosion was also bagged in archival images, constraining the mass of the star that exploded.

Morgan Fraser (University of Belfast) described both the faint explosion and made a detailed attempt at constraining the nature of the star that burst. Most importantly, when confronted by a conundrum, Fraser's team went on to challenge some of the prevailing models for stellar evolution from the perspective of the sorts of underlying processes occurring in the star. The supernova had a peak absolute magnitude of −14, like SN 2008S, nearly an order of magnitude less than other core-collapse events. What was interesting were the archival images revealing a progenitor with a mass of approximately 8.5 times that of the Sun. Admittedly, this in an inference based on the observed luminosity, but the observations point to a star that has a mass less than ten times that of the Sun, but with a luminosity incompatible with a so-called SAGB star.

In essence, here is the round peg in the square hole. SN 2009md was a supernova, it was hydrogen rich and it arose on a star with roughly 8 times the mass of the Sun. However, images imply the progenitor was not an SAGB star. This scenario is ruled out by the luminosity of the star – it was too faint. If the progenitor had been a SAGB star it would have been brighter than observed. SAGB stars obtain the majority of their luminosity by so-called hot-bottom burning, something you can imagine still makes certain members of the astrophysical community titter in seminars. Simply put, this is hydrogen fusion in a shell atop the dense carbon, neon and oxygen core of the giant star. In relatively massive SAGB stars hydrogen burns ferociously at temperatures in excess of 100 million Kelvin – hence the hot part of the term. This occurs in a thin shell and drives vigorous convection all the way to the photosphere of the star. In chemical terms, the process has attracted a lot of attention, as it may be the principle route of formation of fluorine in the cosmos.

Ironically, marginally more massive stars that skip this phase are dimmer than their less massive brethren. Without the prodigious output of hydrogen fusion hot-bottom burning provides, less energy is delivered to the exterior of the star. Observations of the progenitor indicate that it was simply too dim to be a SAGB star and must instead have been a normal, if a relatively lightweight, red supergiant. The only way around this is to imagine that the SAGB star had an extinguished hydrogen burning shell. This is possible if the star was burning helium instead – a process that alternates with hydrogen fusion in evolved red giant stars. Since the phase of hydrogen fusion lasts considerably longer than that of helium fusion, the probability of catching a supernova when the SAGB star was in this phase has to be low, if not exactly negligible. Thus Fraser's team concludes the star was not of the SAGB variety.

The supernova's diminutive luminosity suggested that less than 0.01 solar masses of nickel-56 was produced, again lower than expected for a conventional core-collapse supernova – but perhaps in line with the less energetic electron-capture models.

So, where does that leave all the modeling? In a pickle, probably. Unfortunately, SN 2009md isn't alone. There are a small number of similar Type II-P supernovae that appear to have originated within stars that have lower than expected masses (less than 10 solar masses). They are few in number, but their presence suggests a problem with the underlying models of the evolution of the most massive of the intermediate mass stars. Can stars with masses as low as seven times that of the Sun explode as conventional supernovae? If so, why are they so faint if they produce normal explosions and neutron stars as progeny?

Fraser suggests that perhaps the problem lies with our understanding of the nuclear reactions that drive the star later in its evolution. In particular they focus on carbon fusion. The authors suggest that if this rate is underestimated, the star will complete carbon burning before the core can cool through neutrino emission and become degenerate. This will allow the star to evade such severe mass loss through the third dredge-up and pulsation phases. Henceforth, the star will run through to the formation of an iron core. The scenario is by no means certain, and it is perhaps rather provocative, given the body of work put into SAGB stars in exactly this mass range. Still, a little controversy can go a long way and stir up further research into these troubling stars.

196 Extreme Explosions

A Coda from the Distant Past: Type I.5 Supernovae

We leave what we think we know and enter a more distant realm of proposition and hypothesis. If we change the composition of a star sufficiently can we ignite an altogether different kind of explosion, one that is part core collapse and part thermonuclear in origin? What would happen to a star of modest mass that clung onto its hydrogen layers, retaining them in bulk until the core began to burn carbon under dense, degenerate conditions? Maybe, just maybe, a unique type of supernova might result, the light of which we will likely never see again – a Type 1.5 explosion. Proposed as early as 1969 by David Arnett (University of California, San Diego) and periodically revisited over the decades, Type 1.5 s are genuinely curious explosions. However, it is worth restating the point that if these explosions occurred at all, they were confined to the early universe. They are not visitors on our shores.

Imagine a star born only of hydrogen, helium and a whiff of lithium. Such a star will be hotter, denser and use its fuel faster than a star of today. Why? Because such stars had no heavier elements to soak up radiation from the core and puff them up. Without carbon, nitrogen or oxygen these stars could not use the CNO cycle to burn their fuel. Instead the proton-proton (pp) chain did all the work early on.

If the star was a few times heavier than the Sun – say 5–9 times the Sun's mass – there existed a theoretical route that these metal-free stars could pursue to detonation. During the time the star spent on the main sequence hydrogen fused to helium as normal. But shortly thereafter, the energy supplied by these reactions would have been insufficient to keep the star afloat. The core, still rich in hydrogen fuel, would have shrunk, grown hotter and eventually began to burn the helium at the same time as the remaining hydrogen. At this point, helium fusion would have produced enough carbon to set off the CN cycle and allow a rapid consumption of the remaining hydrogen fuel.

As the core began to collapse again, the star would have swelled into a red giant. Now burning helium in a shell around the increasingly massive carbon-oxygen core the star would have begun to pulse in a manner similar to the AGB stars of today. However, at this point, the evolution of these early stars and those we see around us diverged once more – and this time more spectacularly than our intermediate mass stars.

As we've seen in present-day intermediate mass stars, eventually the star loses all of its hydrogen-rich outer layers through strong stellar winds and the remaining low mass, degenerate core shrivels up as a carbon-oxygen white dwarf. However, mass loss is expected to be considerably lower in metal-free stars, and despite these ancient red giants experiencing significant mass loss, it isn't enough to stop the degenerate carbon-oxygen core growing towards the Chandrasekhar limit of 1.382 solar masses. At this point the star resembles a simmering carbon bomb embedded in a deep bed of hydrogen fuel.

The star never makes it to the Chandrasekhar limit. At 1.36 solar masses the deeply degenerate core can take no more and implodes. The carbon and oxygen fuel begins to burn, generating copious amounts of radioactive nickel-56. A shockwave pushes out into the hydrogen shell, igniting further rounds of nuclear fusion, and the envelope is blown off into space. Buried within the expanding shell lie large amounts of iron, nickel and cobalt, plus a host of lower mass elements synthesized in the explosion. As a result this hybrid explosion generates the radioactivity of a conventional Type Ia explosion (see Chap. 11) with approximately 0.5–0.8 solar masses of radioactive nickel-56. This will mean that such an explosion would have been substantially brighter than any Type II-P explosion seen today. However, this is embedded in a dense outer layer of hydrogen so the supernova has the bright plateau phase of a Type II event and the early spectral characteristics of a Type II event.

Although hypothetical, these supernovae could have contributed some of the iron-group elements seen in the ancient Population II stars of the galactic halo, in addition to those produced by conventional Type Ia supernovae and pair-instability supernovae (Chap. 8). What is needed here is a detailed analysis of the species produced by nucleosynthesis (element production); as well as an honest appraisal as to whether there are sufficiently diminutive stars formed in the early universe that would produce Type 1.5 events.

Conclusion

Although astronomers have predicted the existence of supernovae powered by electron-capture, and elucidated the approximate percentage of supernovae powered by this mechanism, it still remains to be seen if any have been detected. The key criterion for capturing one of these events will undoubtedly not be examination of weak supernovae but the dissection of the ejecta and the precise determination of the route taken by the star to create the observed elements. There is some consensus on what sorts of elements these supernovae should create, and they are rather distinctive in composition. Recent controversy over the origin of weak, calcium-rich, supernovae should, however, serve notice that even with models of nucleosynthesis at hand astronomers don't always agree.

A decade ago, many had the feeling that much of stellar astrophysics had been done and it was simply a case of mopping the floor and shining the silverware. It is now clear that there is much still to do, with many areas of the stellar astrophysics community rife with healthy debate.

8. Ultra-luminous Type IIn Supernovae

Introduction

An older book called *Secrets of Space* (Abacus books) was one of the first I read which split supernovae into different classes and more impressively describe the phenomenon of pair instability supernovae in all but name. It took another 20 years to see mention of this again, such was the uncertainty surrounding their existence. Indeed, they remained theoretical constructs until the discovery of SN 2007bi. Although not confirmed it is likely that this explosion represents the first observation of a type of supernova proposed in the 1960s. This chapter concerns this event and a few other explosions that may be related to this type of explosion.

Taking the Pulse

In Chap. 3 we encountered three unusual supernovae: SN 2006gy, 2006tf and 2008am. All three were of the Type IIn variety, indicating each occurred within dense shells of hydrogen-rich gas. Although the nature of the shells – the dimensions and mass, for example – were somewhat different from one another, the indications were that each was produced by some form of violent cataclysm prior to the death of the star. Each star had shed several solar masses of gas in the years to decades prior to the supernova. The question is, what kind of event could generate such a massive shell of matter in the decades before the star exploded?

In Chap. 3 the phenomenon of luminous blue variables was discussed. These stars are characterized by short, violent outbursts where a fraction of a solar mass or more material is shed in pulses. Very little is known about why these stars behave this way.

However, some of it may be down to unstable energy transfer in their massive hydrogen-rich envelopes, while other sources of instability may lie deeper down in the stellar core. However, a few LBVs take these eruptions to extremes and shed several Sun's worth of mass at a time, brightening until they rival supernovae. Eta Carinae did this in 1843, and this also appears to have happened to the progenitors of SN 2006gy, 2006tf and 2008am, as well as a few other very brilliant supernovae, *immediately* prior to their deaths. (The word "Immediately" is a relative term, in this case meaning years rather than millennia or longer.)

One mechanism that can explain the presence of the eruptions is pulsational pair instability. Although it does not explain the eruptions of all LBVs, including the "great eruption" of Eta Carinae in 1843, it could underlie some of the pre-supernovae outbursts of some of these stars.

Pair instability is a phenomenon confined to stars that have amassed helium-rich cores in excess of 48 times the mass of the Sun at the point they exit the main sequence. The mechanism was first proposed in 1967 by Gideon Rakavy, Zalman Barkat, and Giora Shaviv (Caltech). The explosions caused by this mechanism were subsequently modeled in detail by Gary Fraley (Caltech) the following year. The basic idea is as follows. Massive stars rely heavily on radiation pressure to hold them up against the inward pull of gravity. Should anything weaken it, even slightly, gravity then exerts a heavy toll. In a star such as the Sun, the majority of the pressure that supports the stellar interior comes in the form of particle motion – gas pressure. This is merely an extension of the phenomena that gives rise to air pressure on Earth. At higher temperatures particles have more kinetic energy, moving more quickly from A to B. As they move around they bump into one another, generating pressure.

However, in stars with masses greater than twice that of the Sun, particle motion cannot support the greater bulk of the star. Additional support is needed in the form of collisions and interactions between plasma electrons and photons of light. This radiation pressure steadily increases in strength as the stellar interior becomes hotter, and this increases with escalating mass. In all massive stars, radiation pressure dominates internal support, and without it the interior collapses very rapidly under gravity.

If you recall, at the end of hydrogen burning in a star with only a few times the mass of the Sun, the cessation of nuclear fusion leads to a rapid shrinkage of the core under gravity and the beginning of the red giant stage. The entire process takes less than 50,000 years in a star with 6 times the mass of the Sun. Likewise, in very massive stars, radiation pressure is critical if the star is to remain stable against gravity's fatal attraction. And so, a very massive star lies at the mercy of gravity. If it cannot sustain its prodigious output of energy it is stuffed. Early on, while hydrogen or helium are burning, it's a case of so far, so good. Either mechanism supplies ample energy to support the stellar interior.

However, things get very complicated for all massive stars once helium is exhausted in the core. The next fuel is carbon, followed soon afterwards by oxygen. As we have seen, these fuels are measly contributors of energy per gram compared with hydrogen, so the star needs to burn these at a fair click if it is to produce enough energy to defend its interior against the onslaught of gravity. Unfortunately, as we saw in Chap. 2, carbon and oxygen burning produce a copious amount of their energy in the form of neutrinos. These simply pour out of the star, imparting not an ounce of gratitude for their formation. This is not a terribly helpful state of affairs. Moreover, at the critical temperatures in the stellar core, which exceed one billion degrees – pairs of energetic gamma rays turn spontaneously into electron-positron pairs in accordance with Einstein's eponymous equation, $E = mc^2$.

How does this happen? Positrons are the antimatter opposites of electrons. The electron has a small mass, roughly 1/2,000th that of the proton, and it has a negative charge. It also has a spin in one direction around its axis. Positrons have the opposite: a positive charge, the opposite spin, but the same mass. The two gamma ray photons that give rise to these have no mass, but they do have energy, and if the total energy of these is equivalent to that of the positron and electron, they can turn into them.

Soon after formation, the electron and positron annihilate one another, regenerating the two gamma ray photons. In stars with initial masses of less than 95 times that of the Sun, this process only serves to further accelerate the rate of nuclear reactions, as the star struggles to keep pace with energy losses. This hastens the death-dive of the star, but outwardly, nothing else is notable.

However, very massive stars are so sensitive to even this temporary loss of support that the core becomes fatally destabilized.

In a star with a helium core mass in excess of 48 times that of the Sun (produced by stars with more than 95 times the Sun's mass) pair instability is a crisis. The core loses much of its critical support once its temperature exceeds a billion degrees Kelvin or so. Within seconds the core implodes, and the temperature rises high enough to convert much of the stellar carbon and oxygen to iron-group elements (iron, cobalt and nickel). Remember these elements are normally synthesized by silicon fusion, at a stage later in the star's life. Therefore, the star has executed a significant re-routing of its evolution in the process.

If the mass of the helium core is less than around 130 times that of the Sun, the sudden increase in energy output drives a massive explosion. Now, this is where things get interesting. In 1968 Gary Fraley examined the fates of stars with different masses in the pair instability range. He found that in his models, stars at the lower end of pair-instability range underwent pulsations rather than exploding in their entirety. The surge in nuclear reactions as the core collapsed caused a rapid swelling and outward thrust in the outer core and envelope. Working on this idea and producing more accurate models, Stan Woosley and colleagues found that these stars belch forth a massive wave of material in response to the sudden increase in energy output from the core.

Pair instability releases around 10^{43} J of energy, which shunts the hydrogen-rich outer layers off into space at a thousand or more kilometers per second. 10^{43} J is a lot of energy and is equivalent to a modest supernova, or 17 powers of 10 more than the Sun emits per second. To the observer, this should appear as a rather unpretentious supernova in its own right. However, the star isn't dead. There is more to come.

Rapid expansion of the core causes it to cool swiftly. As temperatures fall below the threshold for pair production, the process shuts off, and the core collapses inwards more. Inward collapse converts the star's massive store of gravitational potential energy to heat, and the core begins to warm up. If the core mass is still in excess of 48 times that of the Sun, the process of pair-instability repeats. Pair production recommences once the temperatures exceed 1–1.2 billion degrees, and the unpleasant binge and purge

Table 8.1 The effect of changing the mass of the helium core on the energy, frequency and number of pulses caused by pair instability. In general lower mass stars have more pulses, as the mass ejected each time is lower and the star is less vulnerable to pair instability to begin with. The longest interval between pulses occurs earlier on, as the stars take longer to adjust to the effects of the pulse. The interval between pulses in the largest model star is over a century, and Nathan Smith suggested this mechanism might account for the massive outburst of Eta Carinae in 1843

Helium core mass (compared to mass of Sun)	Number of pulses	Pulse energy range (multiples of 10^{43} J)	Time between pulses (years)
48	6	0.11–2.4	0.2–0.26
51	4	0.44–3.7	0.09–0.9
52	4	0.94–3.1	0.01–3.0
54	3	2.1–3.2	0.03–12
56	3	1.3–3.3	0.01–110

process starts again. Eventually pulses may reduce the stellar mass below that needed for pair instability. At this stage the core stabilizes and completes oxygen and then silicon burning, as a lower mass star might do. Consequently, the star can complete its evolution with a large iron core (perhaps three times the mass of the Sun). This iron sphere then collapses as normal, forming either a neutron star or a black hole.

Interesting, though this is, it isn't the whole story. Remember that the star may pulse more than once (see Table 8.1) before the core collapses. Now, with each pulse the stellar mass decreases by up to tens of solar masses. Thus each pulse shares out roughly the same amount of energy between lower and lower masses of material. Thus as the mass goes down, the speed at which the material is ejected goes up. The kinetic energy is a product of the mass and velocity of the material. Lower mass means a higher velocity.

From this, some interesting scenarios can play out. Imagine a star with 100 times the mass of the Sun. Its core undergoes a pulse driven by pair-instability. A shell of material, several times the mass of the Sun, is ejected at a thousand or so kilometers per second. The energy in this blast is perhaps 10^{42} J. Thus it produces a very respectable supernova-like display in its own right. However, this is just the beginning. After a brief hiatus the process repeats, and a lower mass but higher velocity shell pursues the first. After a year or so, the two shells collide some distance out from the star. The masses and hence the energies involved in the collision are immense. Several solar masses, perhaps tens of solar masses of material collide, and a large fraction of their kinetic energy is turned into light. The product is a supernova-like display in a broad shell surrounding the still breathing but declining star.

At some time, perhaps a decade or so after the final pulse, the iron core implodes, and the star dies. The brilliant supernova is swamped by thick and possibly dusty shells of material moving outward from the star. This object has already produced two or more supernova-like displays, but soon afterwards the supernova shockwave, moving five or more times faster than the shells, smashes into them, and a final blaze of light is emitted. This time the entire kinetic energy of a supernova (10^{44} J or more) is available to power the display (Fig. 8.1).

Sound familiar? Think back to SN 2006gy and other blasts described in Chap. 3. The principle energy driving these eruptions was the collision between the supernova blast wave and shells of material lying hundreds of millions of kilometers from the star. The scenario described here closely matches the events described for SN 2006gy, SN 2006tf and SN 2008am, among others. Clearly it is possible, perhaps even probable, that these ultra-luminous supernovae, each shining ten or more times brighter than the brightest GRB-associated explosion, is powered by this kind of multiple-punch event – pulsational pair-instability. In general, more massive stars have shorter gaps between pulses, as their cores take less time to reach instability after each pulse. Therefore, we have a testable prediction here. Should we observe several supernova-like displays coming from one star, the frequency and number of pulses should relate to the mass of the star, which in principle is deducible from the total output of radiation. However, determining this may be harder than it sounds… Will all the energy released by the star be detectable by us? One only has to think back to SN 2008S to understand the controversy such observations can elicit.

Potential problems aside, it is worth looking at SN 2006gy, SN 2006tf and SN 2008am again, in the light of the pulsational mechanism, and then decide whether this series of events is the best explanation for these brilliant explosions.

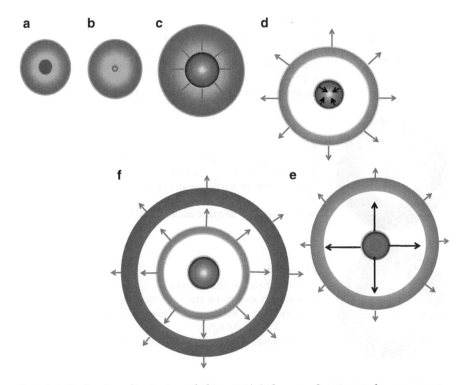

Fig. 8.1 Pulsational pair instability. In (**a**) the star begins to burn oxygen, and electron-positron pairs emerge at high temperatures. In (**b**), lacking support, the core implodes, generating a shockwave. In (**c**) and (**d**) the shockwave expels the hydrogen envelope, and the remainder cools and contracts once more. In (**e**) the core becomes unstable again and generates a second faster moving shockwave. At (**f**) this shockwave catches up with and collides with the first hydrogen-rich layer, generating a powerful supernova-like display

A Lethal Pulse: SN 2006gy and 2006tf as Pulsational Pair-Instability Supernovae

The principle source of light in SN 2006tf came from the collision of the supernova blast wave with an 18 solar mass shell of hydrogen-rich material lying some 160 Earth-Sun distance from the star (2.7×10^{15} cm, 27 billion kilometers). There was clear evidence that the star had ejected 2–6 solar masses of hydrogen in the decades prior to the ejection of this shell. However the shell, itself

206   Extreme Explosions

SN 2008am
Mass of impacted shell: 1 solar mass
Distance to shell from star: Less than 1 billion km
Composition of shell: Optically dense (opaque) hydrogen-rich material
Implied velocity of blast wave: 1,000 km/s

SN 2006gy
Mass of impacted shell: 18 to 20 solar mass shell
Distance to shell: 27 billion km
Composition of shell: Dense but not opaque hydrogen-rich material
Implied velocity of blast wave: 4,000 km/s

SN 2006tf
Mass of shell: 18 solar masses
Distance to shell from star: 27 billion km
Composition of shell: Virtually identical to that surrounding SN 2006gy
Implied velocity of blast wave: 2,000 km/s

Fig. 8.2 Fast facts on supernovae SN 2006gy, SN 2006tf and SN 2008am. The differences in the luminosity of the explosions was due to differences in the energy of the explosion and the distance and properties of the shell of matter surrounding each explosion

moving at 2,000 km per second, appeared to have been ejected in one large pulse, perhaps 8 years before the supernova. Although the shell could have been released in a LBV-like eruption the mass seems unreasonably high. Thus, although not ruling out an LBV-eruption scenario, this seemed less likely overall. Moreover, the release of this shell, 7–9 years before the supernova, and the overall kinetic energy of the shell (7×10^{43} J), are in good agreement with predictions made by Stan Woosley. Recall that the pulsational pair-instability model of Woosley, Bilinkov and Heger was proposed for SN 2006gy for a star with 110 solar masses. Although not constrained by observation, Woosley's PPI model for this mass of star matches the properties of SN 2006tf and its environment particularly well. However, this isn't proof; it's just a good match to a model (Fig. 8.2).

What is interesting about this explosion is the implication that the outgoing shock wasn't particularly affected by its collision with the shell. The velocity of the shock appeared to be fairly constant, implying that the explosion must have been particularly energetic. The light curve allows for the presence of 4.5 solar masses of radioactive nickel. One must remember that this is just a correlation, not a definitive cause. The light curve allows but does not prove the existence of this mass of nickel-56. Moreover, whether a conventional core-collapse supernova would generate this (large) amount of radioactive nickel is certainly unclear.

Similarly, the light curve of SN 2006gy allows for a contribution from the decay of 2.5–3 solar masses of nickel-56. However, the overall luminosity can only be accounted for by 22 solar masses of this element – a vast and unreasonable yield for even a bona fide pair-instability explosion (see later in this chapter). Clearly there is a schism between what the light curve suggests and what the overall output of the supernova demands. Therefore, the contribution of radioactive nickel to the light curve of these supernovae is questionable, to say the least. In these instances adding a contribution to the supernova's glow from the decay of nickel-56 almost seems apologetic.

As with many observations of supernova the problem with SN 2006tf is the lack of data from the early part of the supernova. The first observations may have been made as late as 70 days after the explosion occurred, and hence there is little to constrain what happened during this first phase of the eruption. This, in turn, limits what information can be gleamed from the light curve. Not a catastrophe, but it certainly necessitates a note of caution.

Let's turn then to SN 2006gy and SN 2008am, for which there is an abundance of early observations, particularly in the former's case.

SN 2006gy was observed in the central regions of a spiral galaxy with near solar metallicity. The explosion was characterized by its brilliance, its relatively slow rise to maximum light (70 days) and its prolonged decline over the course of more than a year. There was clear evidence for the presence of a thick, high-mass shell of opaque material lying around the star; and it was the collision of the supernova blast with this shell that powered the majority of the output of the explosion.

The length in the delay to reach maximum light implied that the surrounding dusty shell lay some 10^{15} cm or so from the star – a very similar distance between the star and shell in SN 2006tf. Likewise, SN 2006gy appeared to have shed its 10–20 solar mass shell 8 years or so before the supernova. This was yet another area of similarity between the two events. Compared to this lesser explosion, SN 2006gy appeared to be more energetic, with a blast wave moving at a speed twice that of SN 2006tf. The extra oomph in the shockwave largely explained the difference in the brightness of the explosions: more kinetic energy in the shockwave of SN 2006gy meant more energy was available in the collision. Furthermore, late spectra indicated the presence of another more distant, 10 solar mass, dusty shell. This implied that SN 2006gy's progenitor had made something of a habit of such ejections; this first shell having been discarded perhaps 1,000 years before the supernova.

By 200 days after SN 2006gy a 10 billion kilometer-thick shell, containing a total of 18–19 solar masses of hydrogen-rich gas, had been swept up by the blast wave. Spectral analysis indicated that 50 % of the kinetic energy had been converted to visible light, powering the supernova's light curve. This massive shell of material appeared to be embedded in a broader but optically thin (largely transparent) zone of circumstellar material, perhaps 100 billion kilometers thick. Thus the light from SN 2006gy illuminated a considerable episode in the life of its progenitor star as the supernova stumbled towards its annihilation.

Of interest is the coincidental observation that Eta Carinae has also pulsated over this time frame. Lying distant to the currently expanding homunculus is a much more extended shell of matter that appears to have been ejected 1,000 years ago. Is Eta Carinae on route to a pulsational pair-instability outburst? Therefore, like both progenitors of SN 2006gy and 2006tf, Eta Carinae has shed massive shells of matter in the run-up to its eventual supernova.

What of supernova SN 2008am? This, as it turns out, is a slightly different beast. Although marginally brighter than SN 2006gy, prior to its conflagration the progenitor seems to have undergone a milder eruption than either SN 2006gy or SN 2006tf. SN 2008am had an estimated energy of 2×10^{44} J, with a peak absolute magnitude of –21.7, making it the second brightest supernova on record at the time of discovery. The brilliance of the event implied a very efficient conversion of kinetic energy into light in this explosion. The 34-day rise time to peak light was also significantly shorter than that of both SN 2006gy (70 days) and the poorly constrained rise of SN 2006tf (perhaps 50–100 days). Based on its peak absolute magnitude, this spectacular light show would require 19 solar masses of radioactive nickel-56 to power the display and provide the energy in the moving matter.

The blast wave of SN 2008am had a fairly modest speed of 1,000 km per second – half that of SN 2006tf. What was significant was the mass of the shell – perhaps only 1 solar mass. However, in order to power the greater peak display, the shell had to lie much closer to the star at the time of the supernova. This meant that the supernova shockwave intercepted it at an earlier time, and consequently the early interception meant that much more of the kinetic energy in the supernova shock wave was converted to light. It also seemed likely that the less massive shell of SN 2008am was very dense, allowing little of the light generated by the collision through until the shock had traversed and thoroughly heated the entire shell. This would also ensure a much more dramatic pulse of light once the shockwave had passed through, hence the higher peak absolute magnitude.

What was the origin of this small, dense shell? SN 2008am's progenitor appeared to have suffered a smaller eruption, perhaps a few years before the supernova. The ejected shell was similar in mass to that ejected by the LBV P-Cygni in its 1600 eruption or by the progenitor of SN 2005gl. Almost certainly, a PPI explanation is unwarranted in this case, and a standard LBV-eruption could account for the properties of this shell.

However, the substantially larger masses of the shells surrounding both SN 2006gy and 2006tf seem too substantial to be accounted for by a simple LBV eruption, and the timing of the event, immediately prior to the supernova, also seems like too much of a coincidence. However, with only a handful of such events a simple coincidence wouldn't come as such a shock.

The only definitive test of the PPI model for these explosions is to catch one or more of them exploding again. In the case of both SN 2006tf and 2006gy, the first, prolonged eruption may simply be a gloriously extravagant imposter; the final demise of the star in a core-collapse event may yet lie in the future. In this case, we'll see yet another spectacular rise in brilliance of the currently fading remnant as the star explodes, and the blast wave catches up and incinerates the current outgoing shell of matter.

Independently, we could see the entire process of pulsational pair instability unfold in a separate event. A distant star is seen to brighten dramatically, followed a few years later by an even more brilliant explosion. The first burst demarcates the star's first pulse, and the second more spectacular event the next, as the second outgoing shell overtakes the first. Clearly to make such observations requires more than a little luck as well as a heap of ingenuity. There is a very large volume of space to sample with current technology.

What is remarkable about these supernovae, whatever their ultimate origin, is the mechanism powering their output. A typical Type II supernova derives most of its output from the decay of nickel-56 (Chap. 2). There is a contribution from photospheric expansion and heating as the shockwave moves through. However, the bulk of the energy observed on Earth ultimately is derived from the radioactive decay of nickel-56.

If we consider the majority of Type IIn explosions – i.e., those showing evidence for interaction of the blast wave with surrounding material – once again, most of the light comes from radioactive decay. Type IIn supernovae show the effects of flash ionization (the source of the narrow lines in the spectra) and the effect of the blast wave that has interacted with and heated the surrounding hydrogen-rich material, generating intermediate or broad lines (the width dependent on the speed the that the shockwave is moving at).

Even with their unique traits, the generation of visible light in most Type IIn explosions follows an intrinsically similar route to the Type II-P events. However, in these ultra-luminous Type IIn explosions, the material surrounding the stars is extremely dense, and the shockwave suffers violent deceleration as it punches through the material. These supernovae are powered not through the decay of radioactive elements but by the conversion of kinetic energy to light. Thus they are quite unlike any other supernovae in this book.

Even when we consider the broad-lined, GRB-associated, Type Ic supernovae the energies involved don't quite measure up to those seen here. Where the broad-line Type Ic events fall down is the limited scale of the jet. With an opening angle less than 10°, the scope for interaction with the star or any surrounding material is small. Thus the energy the opposing beams impart to the star is limited. Consequently, the supernovae are not unusually brilliant – at least not by the high standards set here.

SN 2008es: An Ultra-Luminous Type II-L Supernova

This is an enigma, plain and simple. Like the brethren considered before, this event was apparently powered by some form of collision. Sounds fine, but – and it's a big but – it's a collision for which there is no direct evidence for any surrounding shell. Confusing? Perhaps.

SN 2008es is worth mentioning if only for the fact that it outshone all the other supernovae thus far discussed. At a peak magnitude of –22.3 it was exceeded only by the enigmatic SN 2005ap and SCP 06 F6 (see Chap. 10). The total energy radiated by the supernova over days 21–75 was equivalent to that emitted by SN 20006gy over its first 2 months.

What sets this supernova apart from the others in this chapter is the shape of the light curve. The others display ample evidence of blast wave-circumstellar interaction in the form of narrow- and intermediate-width emission lines. These can only be produced through collisions between shells of gas. SN 2008es showed none of this.

Instead SN 2008es showed a relatively rapid rise to maximum light and a concomitantly fast decline (0.042 magnitudes per day). The overall shape of the light curve matched so-called Type II-L supernovae – one where there is no plateau phase in the declining light curve. These supernovae are often explained as conventional Type II explosions that occur in stars with meager hydrogen envelopes. In Type II-L supernovae, the material is shocked, heated and irradiated as normal, but with little hydrogen, the expanding cooling debris lacks the plateau caused by the recombination of hydrogen ions with their electrons (Chap. 2). The light curve is thus dominated by the decay of nickel-56 and its daughter, cobalt-56.

The light of SN 2008es certainly indicated the presence of hydrogen, and the light curve, although linear, was a poor match to the decay of nickel-56. The overall luminosity for the supernova demanded a yield of nickel-56 on the order of 10 solar masses, but this was not matched by the shape of the decaying light curve. Once again, the source of the supernova's brilliance appears to be a collision. The question is, what sort of collision generates a Type II-L light curve without narrow- and intermediate-width emission lines? Moreover, radio observations failed to identify the explosion. Although many Type IIn explosions are radio loud, there are a few that are radio quiet. Perhaps, then, this feature should not then be used as a factor to discriminate against a collision.

Andrea Miller suggested that this brilliant Type II-L explosion might be explained as follows. A yellow hypergiant star similar to rho Cassiopiea has a mass of around 30 times that of the Sun (Chap. 3). A few years before annihilation the progenitor of SN 2008es had been a red supergiant, but it was now undergoing a blue loop in its evolution. The stellar radiation had begun blowing away the cobwebs of its former life. An old, dusty red supergiant could no longer hang onto its envelope; it was being driven away by the intense stellar radiation. As the star evolved it began to carve out an expanding cavity in the thick, dusty stellar wind. The star was now shrinking, heating up and evolving into a blue supergiant.

Unfortunately, death overtook the star before it could complete its evolution. The star detonated inside the cavity it was producing in its opaque shell of gas. All of the ultraviolet light from the shock break-out was then absorbed by this shell of gas. Therefore, there were no narrow emission lines in the spectrum of the supernova. A few days after the detonation, the shockwave had traversed the 20–30 billion kilometers gap around the star, arriving at the inner edge of the shell. The shock compressed and heated the dense shell, causing a sharp increase in optical luminosity. Behind the shockwave the star's remains expanded but were relatively hydrogen-free, the star having shed most of this layer in the previous years. Thus, the subsequent decline in the light curve was linear, as most of the hydrogen already lay at a great distance from the star in the shell. Once the shock emerged from this layer there was little further hydrogen-rich gas at greater distances. There was nothing left to energize and generate the narrow- and intermediate-width emission lines characteristic of Type IIn events. Thus the Type IIn phenotype was avoided, and the explosion appeared enigmatic.

In Miller's scheme the supernova's brilliance was apparently a consequence of a collision that was hidden from the spectroscopes of Earthly astronomers. If the spectroscopy was ambiguous, was there anything else that supported the assertion that the supernova was collision-powered? Although narrow and intermediate-width emission lines were absent, spectroscopy did reveal one clue – the evolving temperature of the ejecta. Early spectra indicated that the gas in the vicinity of the supernova had a temperature of 15,000 K, but that this fell progressively to 6,000 K over the course of the observations. Miller and colleagues could then infer that they were observing a retreating photosphere: the *surface* of the supernova's light-emitting region as it shrank inwards through a very extended layer of hydrogen-rich material. However, unlike the case in a conventional Type II-P event (see Chap. 2), the gas was very opaque and lying distant to the star, and thus the radiated energy was absorbed. This allowed the supernova to avoid the plateau phase of Type II-P events but still retain the imprint of the hydrogen-rich shell.

To summarize the conclusions: the progenitor of SN 2008es was likely an extended yellow supergiant star, with a much lower mass than the progenitors of SN 2006gy or SN 2006tf. The inner edge of this shell lay 20–30 billion kilometers from the star, roughly 10 times the distance to Uranus or Neptune, respectively. The star had exchanged most of its envelope for this thick cloak, which it now wrapped around itself. When the star exploded the ultraviolet light, accompanying the shock break-out, was absorbed by this extended veneer of material. Absorption of the shock's ultraviolet pulse prevented the formation of the visible, narrow emission lines in the spectrum and effectively hid the circumstellar material. As the shockwave passed through, the emitted light peaked as the shock neared the edge of the shell, then declined linearly. Subsequently, the photosphere of the supernova retreated back inside this dense shell as the temperature of the gas subsided. Although somewhat contrived, the scenario is not unreasonable and stitches with the evolutionary models of stars in the 30–40 solar mass range.

The key difference between SN 2008es and other supernovae was the opacity of the shell. The shell surrounding SN 2008es was more opaque, and little light could leave it until the shockwave had traversed its full extent. This then links the proposed scenarios for SN 2008am and SN 2008es, except that the shell surrounding the former was also embedded in a more extended, hydrogen-rich outflow. Conversely, the shells around SN 2006tf and 2006gy were so massive, dense and extended that they effectively captured the energy in the shockwave and progressively converted it to light. The supernovae's light then rose as the shockwave swept up and energized this. But because they were less opaque to visible radiation than SN 2008es, the shells released more energy early on, causing these supernovae to peak more gradually. Finally, like that of SN 2008am, the shells surrounding the progenitors of SN 2006gy and SN 2006tf were also embedded in a much more extensive outflow of hydrogen-rich matter. Upon exiting these shells, there was still ample hydrogen-rich material to energize, generating narrow and broad emission lines. Thus SN 2006gy, SN 2006tf and SN 2008am were Type IIn events, while SN 2008es was a linear one.

In turn, the data suggested that the progenitor of SN 2008es was lower in mass than the other supernovae discussed previously. There was clearly less material deposited around the star prior to its demise. To reiterate, the stellar progenitor of SN 2008es was more like the 40 solar mass rho Cassiopiea, rather than an Eta Carinae, weighing in with 100 solar masses. Thus ironically the latter scenario is necessitated by the dimmer explosions of SN 2006gy and SN 2006tf, rather than SN 2008es. The light curve of SN 2008es compares well with the more luminous Type Ic SN 2005ap (Fig. 8.3). Although relatively little is known about the latter event (Chap. 10) the similarity in the events may point to a similar mechanism powering their extreme luminosities.

An alternative, but not exclusive, scenario invokes some form of jet. In some highly luminous Type II-L events astronomers have

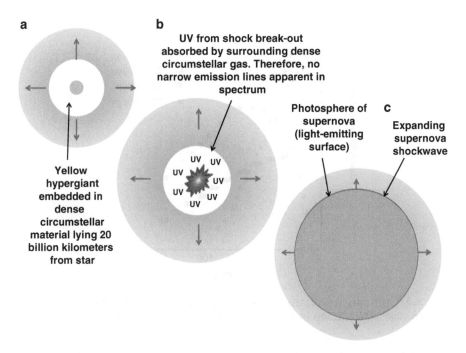

FIG. 8.3 The proposed mechanism of Miller et al., which may explain the peculiar but highly luminous SN 2008es. A yellow hypergiant emits a dense, opaque wind that completely encases the star (**a**). The star then detonates, and ultraviolet light from the shock break-out is absorbed by the hydrogen-rich shell of gas (**b**). When the shockwave emerges from the edge of the dense shell the light peaks then decline rapidly, as there is little further material for the shock to heat. The supernova photosphere then retreats back inside the shell as the gas cools

suggested the extra luminosity comes from the effect of viewing a jet beamed at the observer. Although there is no evidence for an accompanying GRB or X-ray flash, it is not inconceivable that the supernova had some form of asymmetry. Perhaps a jet was submerged in a thick sea of material – either the star's outer layers and/ or the proposed outlying shell of material. Either way any jet from the collapsing core could have had most of its energy absorbed and redistributed by the circumstellar material. This energy was then re-radiated as visible light (Fig. 8.4). What happened was a Type II-P event, but where the steady torch of the plateau phase was being drowned out by the extra luminosity of the jet (Chap. 1).

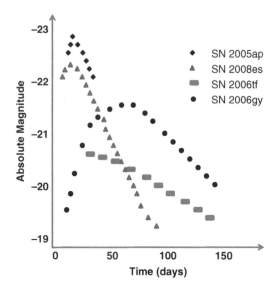

FIG. 8.4 Comparison of the light curves of four of the most luminous supernovae. SN 2005ap is discussed in detail in Chap. 10

Whatever the explanation, it suggests that ultra-luminous supernovae come in different guises. SN 2008es apparently had the same underlying mechanism as SN 2006gy and several others. However, the geometry of the surrounding shell of matter was different from that found in SN 2006gy. Therefore, the shape of the light curve differed.

What we do know is that these ultra-luminous supernovae are very rare – perhaps ten times less common than the supernovae associated with long gamma ray bursts. They may occur at less than three explosions per ten million supernovae per cubic megaparsec every year. This is a very large volume of space, a volume equal to a cube with over 3 million light years along each side. Therefore, the circumstances needed to produce them must also be rare. They may represent the deaths of the most massive and hence rarest stars – SN 2008es excepted, or it may take particularly uncommon circumstances to produce these stars, such as stellar collisions in dense, high-mass star clusters (Chap. 6). There certainly seems little to tie them to unusual metallicity. SN 2006gy had near solar metallicity, while SN 2008am was probably metal-deficient. Apparently, new paradigms are needed. Simon Portegies-Zwart proposed that a stellar collision may have synthesized the unusual progenitor of SN 2006gy: this might just do the trick for other ultra-luminous events and explain their rarity (Chap. 6). Whatever the route, in general these stars appear to be particularly massive. The implication is that some of them die through novel mechanisms, perhaps pulsational pair instability rather than the conventional core-collapse route.

A Deadly Couple Embrace: Was SN 2007bi the First Pair Instability Supernova?

It is a tiny pinprick of light lost in a sea of granulation. When you first look at the raw images of SN 2007bi (or indeed many other distant events), it isn't much to look at. However, the banality of the image belies a rich discovery. Data would soon suggest that this little speck of light was the first detected pair instability supernova.

As we have already seen stars with an initial main sequence mass in the range 95–130 times that of the Sun are believed to produce pulsational pair-instability events. Yet more massive stars are expected, in theory at least, to completely self-destruct in more vicious pair-instability explosions. However, theorists are confronted with the thorny issue of mass loss through stellar winds.

If some stars can cling onto 130–260 times the mass of the Sun they might just expire through this dramatic mechanism. As it isn't clear how much mass stars lose, astronomers prefer to refer to the mass of the helium core as the true measure of stellar fate. Therefore, we will look not at the whole star but at the helium core as a more precise indicator of stellar mass at the point the star expires.

In theory the estimated mass of the helium core in a massive star is about half that of the star, as a whole, at birth. A 95 solar mass star has a helium core on the order of 40 times the mass of the Sun, and one with an initial main sequence mass of 130 solar masses has a helium core with a mass of 60 times the mass of the Sun. Using the helium core mass is more useful, particularly because mass is constantly shed from the outer layers of the star, making the final mass at death difficult to pin down.

SN 2007bi was a Type Ic explosion, signifying that the progenitor had shed its helium and hydrogen layers, either through stellar winds, violent pulsations or to a nearby companion star. Estimates of the mass of the stripped remnant put the original helium core mass at least 100 times that of the Sun. However, with hydrogen and helium shells missing, the progenitor must have been significantly more massive at birth, probably weighing in at more than 200 times the mass of the Sun. In turn the radioactive nickel yield was best matched by the death of a 100 solar mass helium core, and the observed light curve made a good match for this model. If one assumes that these assertions were correct the only suitable stellar model would have been a pair-instability event – one in which pair instability has destroyed the entire star.

So, how can pair instability completely annihilate a star? In stars with masses of more than 130 times the mass of the Sun, the helium core is so large at the end of hydrogen fusion that it embarks on a short, violent trip towards obliteration by pair instability. The helium core weighs in at more than 60 times the mass of the Sun. If mass loss is high while the star is burning hydrogen, the entire hydrogen envelope is shed before the star leaves the main sequence. The star evolves into a WR star. The effective mass of the core then continues to shrink while helium is fusing to carbon and oxygen and mass is blasted away into space on strong stellar winds. In principle this could help a star avoid pair instability. However, we already know that some stars appear to retain more than 100 solar masses of material up until they die.

Any star that retains a helium core with between 60 and 130 times the mass of the Sun is expected to die through pair instability. Obviously, stars become rarer as their mass increases. Much like bubbles in the bath, the largest bubbles are the rarest and lead the most fleeting existences. Therefore, catching their deaths is akin to winning the lottery a few times in a row. The fraction of supernovae that are pair-instability events is likely to be far less than one in 10,000, but in reality any figure is highly uncertain.

The doomed star is a brilliant beacon, shining with the light of several million Suns. As it evolves off the main sequence little is likely to be left of its hydrogen-rich envelope, even if it began life with a few hundred solar masses of hydrogen. By the time a further 100,000 years has passed the helium core is exhausted, and contraction and heating leads to the ignition of carbon. Still unstable, the core continues to contract and heat more. At a value close to one billion Kelvin and on the point of oxygen ignition, pair instability strikes.

The core begins a rapid, violent descent. The temperature rises to over three billion Kelvin, and carbon and oxygen begin fusing to make iron. Still the energy is not enough to halt implosion, and the rate of nuclear reactions accelerates. At some critical juncture, seconds into this calamity, the cabal of particles produces enough energy to halt collapse and drive a violent outward explosion. The energy is sufficient to completely unbind the core of the star and drive the remaining envelope out into space. In the process several solar masses of radioactive nickel-56 are synthesized, and the star dissolves into a fractious mass of expanding debris.

The large stellar mass and concomitant strong gravitational pull slow the expansion to the observed values of around 5,000–6,000 km per second. As the debris expands the large yield of nickel-56 (5–10 times the amount produced in thermonuclear supernovae) powers a long-lasting display. What's important is that the light curve is driven in almost its entirety by radioactive decay. There is likely some contribution from collisions between the expanding debris and surrounding stellar winds, but this is probably minor in effect. With a very large yield of nickel-56 under its belt the supernova takes 70 days to rise to peak brilliance. The rise time is comparable to that of the Type IIn supernovae SN 2006tf and 2006gy – if considerably less than the exceptional SN 2008iy.

What is unique here is the prolonged decline, lasting over 500 days. The sluggish subsidence in luminosity, amounting to 0.01 magnitudes per day, is a clear indication that an awfully large amount of radioactive nickel-56 has been synthesized in the explosion.

We have already seen that both SN 2006tf and 2006gy could, in principle, have been powered by comparable yields of radioactive nickel. Yet this explanation was largely disfavored by the astronomical community. So, why accept this yield of nickel for SN 2007bi? The answer was two-fold. First, we had clear evidence that SN 2006gy and 2006tf were powered largely by collisions of shells of gas. The evidence is abundant in their spectra.

The spectrum of SN 2007bi showed no evidence for this kind of explosive interaction. Secondly, the light curves of both SN 2006gy and 2006tf were matched by declining energy from the conversion of kinetic to light energy, as were the sluggish expansion speeds of their ejecta, compared to other supernovae. However, SN 2007bi had a light curve that reproducibly matches the decay of nickel-56. Thus we were left with a supernova that seemed to be powered largely, or perhaps exclusively, by the decay of a large mass of radioactive nickel.

Comparison of the supernova's light curve to computer generated, or synthetic, curves suggested that between 3 and 10 solar masses of nickel was likely produced in the explosion (7 solar masses was the best match). On these grounds alone, SN 2007bi was likely a genuine pair-instability event. However, when you superficially examine the data you might question this assumption. If SN 2006gy was a pulsational event, the weaker cousin of a full blown decimation of a star by pair instability, wouldn't the mightier sibling be both brighter and longer lasting? SN 2007bi certainly wasn't brighter. SN 2006gy came in at an absolute magnitude of −22. By contrast SN 2007bi was a bit of a lightweight, coming in at −21.3, or roughly three times dimmer. However, all things are relative. Remember, although SN 2006gy soon lost its crown as the most brilliant supernova, it was still orders of magnitude brighter than a typical core-collapse supernova. Likewise, at −21.3, SN 2007bi easily outshone all the identified GRB-supernovae at that point and SN 2006tf. Moreover, if there were questions as to this supernova's ability to acquire the crown for top-dog in terms of sheer brilliance, it more than made up for any deficiency with its sheer staying power. The total radiated energy lay in the region of $1-2 \times 10^{44}$ J, particularly high for a supernova. The explosion's kinetic energy was also high at 100 times this value, and taken together the overall energy was 100 times that of a conventional supernova. Even SN 2006gy is dwarfed by the sheer energy of this event. The promiscuous term "hypernova" could certainly be applied here.

What else? Although not very well constrained, the spectra revealed a bipartite distribution of speeds of the ejecta. A fraction of the material had a very high velocity – 12,000 km per second – while the bulk of the ejecta had a much slower velocity of

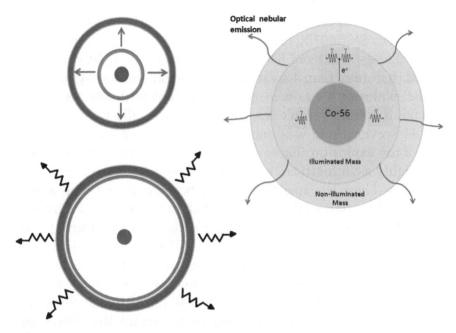

Fig. 8.5 Pulsational pair instability (*left*) versus pair instability (*right*). In the former, collisions between shells of matter generate abundant radiant energy. In the latter, the radiated energy comes directly from the radioactive decay of copious quantities of cobalt-56 (the daughter nucleus of nickel-56)

5,600 km per second. It is tempting to think that the fast-moving material may have been in the form of jets. However, there was no evidence in support of this. Observations of the supernova were unfortunately sparse in early times. The slower-moving material might be expected in the explosion of such a massive star. Much of the material's kinetic energy would be sapped as it moved outward through the star's gravity well (Fig. 8.5).

Were there any cousins of SN 2007bi at the time of its discovery? If we exclude the possible, yet improbably high, yields of nickel that could be inferred from the light curves of SN 2006gy, SN 2006tf and SN 2008am, we are left with only one supernova that showed a reliably, and comparably, high yield of radioactive nickel, the Type Ic explosion, SN 1999as. So, was SN 1999as a pair-instability supernova?

Surprisingly, relatively little has been published about this highly unusual but very brilliant supernova. It had high peak luminosity with an absolute magnitude of −21.4. This is almost identical to SN 2007bi, exceeds SN 2006tf (−20.7) and comes close to SN 2006gy (−22) and SN 2008am (−21.7). The light curve suggested that more than 4 solar masses of nickel-56 was synthesized in the explosion – again in good agreement with SN 2007bi. At this point you may think yes, these are essentially twin events.

However, there was one rather curious and interesting feature of the explosion. There was clear evidence that the explosion was highly asymmetrical, generating two opposing and very energetic jets. These jets appeared to have been generated deep within the exploding star, as they contained abundant iron and moved at 12,000 km per second, or 4 % of light speed. The supernova seems very similar to the "hypernovae" SN 2002ap and SN 2003jd. Both of these explosions showed evidence for fast-moving ejecta directed in opposing jets. However, the evidence for asymmetry was far more equivocal in the spectra of SN 2007bi. Although there was evidence for a faster-moving component within the debris, the bulk of the material was not particularly fast moving nor was the orientation of the faster component certain.

However, such high yields of nickel-56, and the presence of opposing jets, are predicted in some core-collapse models by Naoki Yoshida and Hideyuki Umeda (Both at University of Tokyo), which followed on from earlier work by Ken'ichi Nomoto (also at the University of Tokyo). Indeed, the spectrum of SN 1999as was rather similar to that of the well-documented GRB-supernova, SN 1998bw, except that the yield of nickel was an order of magnitude higher. Although there was this difference in the yield of radioactive nickel-56, the other features of the spectrum were qualitatively similar. Therefore, was SN 1999as a very energetic core-collapse event rather than a pair-instability explosion, and thus could SN 2007bi be explained with a similar model?

Aside from the underlying yield of nickel-56, the comparison with SN 1999as only serves to muddy the waters. The two supernovae appear to be rather different (the data is unfortunately sparse and it is therefore difficult to make a reliable comparison). We need to catch a similarly fecund, and nearby, supernova while its light curve is in its rise phase. At this point we should be able to discern the effects of jetting. Jets reveal themselves through changes in the slope of the light curve between early and mid-late phases. From this data we can decide whether the bulk of the energy is present in two opposing jets, as would be suggested by the core-collapse model, or whether it is spread more uniformly in an expanding fireball powered by pair-instability. Early spectroscopy of SN 2007bi would have revealed this, and with some luck a future event may give up this secret.

A pair-instability event will brighten in accordance with the production and shedding of nickel-56, whereas a core collapse jet explosion will show a number of other features, such as strongly polarized emission caused by fast-moving jets of material. Once you reach the peak luminosity the light curve of both may be dominated by the decay of nickel-56's daughter nucleus, cobalt-56, in both situations.

The problem with the observations is they don't cover enough of the rise to clearly differentiate between both models. As Ken'ichi Nomoto points out, either model works for SN 2007bi. What favors the pair-instability model is the clear similarity between the modeled light curve of a 100 solar mass helium core and a pair-instability event, with the light curve of the explosion itself.

Were there any holes in the data that could cast doubt on the pair instability interpretation? The only real fly in the ointment, and one that is readily explainable, was the apparently lower than expected mass of debris picked up in the nebular phase of the explosion. Within this there was a lack of some lighter elements in the supernova spectrum that would have been expected to have been produced in the pair-instability event. The authors linked both discrepancies to the properties of the ejecta, which were assumed to block some of the light from radioactive decay so that they were unable to illuminate the "missing" material. This is a reasonable, if not entirely satisfying, explanation for the spectral peculiarities. We will need to await further, suitably close relatives to really tie down the pair instability death mechanism.

While we await the destruction of a nearby humungous star, Jeff Cooke (Swinbourne, University of Technology) and colleagues identified two further super-luminous supernovae (SLSN) at great distance using archival images. SN 2213-1745 and SN 1,000+0216 were identified in stacked digital images of deep-field exposures taken by the Canada-France-Hawaii telescope. Buried within these enhanced images were two distant supernovae at red-shifts of 2.05

and 3.90, each with a peak absolute magnitude of approximately −21. Although not record-breakers, these very distant events were clearly unusual. The first of these, SN 2213-1745, bore strong similarities to SN 2007bi and appeared to be a distant analog of the prototype pair-instability event. Like SN 2007bi, this explosion was hydrogen-poor, and its slowly declining light curve appeared to be powered by the decay of a large mass of radioactive nickel-56. Cooke and others estimated that these nickel-powered supernovae appear to be a tenth as common as other SLSN, which are perhaps in turn as rare as one in 1,000 supernovae. Think of very small, brown needles in very large haystacks.

The second SLSN discussed by Cooke was SN 1,000+0216. Although this event was not as bright as SN 2006gy, the rise and fall time of the explosion was similar, implying a common underlying mechanism. The larger overall luminosity and longer rise in the light curve suggested the outburst was caused by the collision of shells of matter around the central explosion. Although the data was relatively sparse, Cooke and colleagues suggested that these distant analogs of pulsational and pair-instability explosions may represent the tip of a larger iceberg lurking in the distant cosmos. Back-of-the-envelope calculations by Cooke suggested that the rate of these SLSN events is ten times greater at red-shifts greater than 2 billion (or more than 8 billion) years ago than in the present universe.

If the technical problems associated with obtaining usable data from these distant events can be overcome, we may be able to determine how these most rare of beasts live out their lives in the wild, even though they make only the most fleeting of appearances in our backyards.

One must remember that the models of pair-instability explosions are still rather crude. Effective modeling of rotating stars is needed. Can a pair-instability explosion generate the strong asphericity seen in many Type II explosions or even organize well-defined, high velocity jets? You could envisage the core collapsing into a rotating disc as pair instability gets underway, and the explosion then launching jets or broad outflows perpendicular to this disc. However, such a "gedanken experiment" is just that. It's a skeleton in need of flesh and blood; most importantly it requires validation through observation. The real beauty of the current astronomical picture is that there are so many automated, high throughput supernova searches running that such observations are likely to come sooner rather than later – even if these pair-instability explosions are as exceedingly rare as Cooke and others suggest.

Was Pair Instability a Common Cause of Death in the Early Universe?

According to some theories the first stars were likely to have been particularly massive. Why? Because, as was described in Chap. 2, low metallicity is thought to result in poor fragmentation of the cloud cores that give rise to stars. Therefore, these cores remain massive, and subsequently the stars that are formed are massive as well.

Yet, some of the pillars of these models are being chipped away at from different directions as a consequence of observations. The models of the early universe suggest that very few low mass stars could form. However, stars have now been found dating back to shortly after the Big Bang that have metal contents as low as one millionth that of the Sun.

Just how small can a star be if its metallicity is very low? Do the first stars birthed in our universe have to be as enormous as we presume?

Approaching the same problem from the other end of the mass spectrum, pair-instability supernovae should produce particularly iron-rich debris, with a well-defined pattern of isotopes. Core-collapse supernovae, by contrast, tend to produce material much richer in oxygen, magnesium and other intermediate mass elements. If pair-instability supernovae were common in the early universe, we should see their fingerprint in the chemistry of the subsequent generations of stars, and we don't. However, the chemotype of the oldest stars we observe is clearly dominated by the imprint of core-collapse supernovae; with additional contributions from thermonuclear supernovae and stellar winds from lower mass stars. The implication is that if there were pair instability explosions early on, these were relatively rare. The majority of supernovae were conventional core-collapse events.

Moreover, modeling work by Sylvia Ekström, George Meynet and subsequently, Cristina Chiappini and André Maeder (all at the Geneva Observatory, Switzerland) suggested that the earliest stars were likely fast spinners. These "spinstars," as Maeder and colleagues called them, would readily shed mass through powerful equatorial stellar winds. Something like centrifugal force would have allowed significant mass to be shed from these initially very massive objects, until they dropped below the threshold for pair instability. They would then have exploded as conventional core-collapse supernovae. Although this part of the universe's early history is effectively lost to us, the chemistry of these spinstar supernovae certainly matches the observations of many astronomers, Ken'ichi Nomoto included, and would help solve the puzzle of the earliest stars.

We ask once again, was SN 2007bi a pair-instability explosion, and if so, why did this star die in this manner while so few others preceded it? The first question is difficult to answer conclusively, as the observations of the supernova were incomplete. However, what we can say is that pair instability is probably the best explanation for the data we have. Ken'ichi Nomoto's alternative explanation that the explosion was a standard, if very powerful, core-collapse event works but rests on data we do not have – the spectra of the earliest stages of the explosion. Low metallicity, so often the first port-of-call as an explanation of strange stellar events, is probably a red herring.

There was a famous problem with globular clusters. Some had their older population enriched in blue horizontal branch (helium-burning) stars, while others were redder. In attempting to solve this problem metallicity was initially approached; however, the data just didn't add up. Nor does it here. SN 2007bi was hosted by a galaxy with a lower proportion of metals than the Sun, although not greatly so. SN 2006gy was definitely not metal-poor.

Perhaps, then the progenitor of SN 2007bi was a slow rotator – simply able to cling onto its initial mass, or perhaps its mass was boosted through stellar mergers, collisions or accretion from a companion. Such aggressive encounters might also explain both the progenitor's high mass at death and the lack of hydrogen or helium in the spectrum of the explosion. In this scenario the latter two shells were ejected in violent merger events. Stellar collisions or harassment also appear to be the solution to the variation in color in the globular clusters that was referred to earlier. In denser clusters, more red giant stars are involved in hit-and-run incidents with other stars, which disperses their hydrogen-rich outer layers. Perhaps it was a combination of lucky circumstances that bequeathed the progenitor of SN 2007bi with enough bulk to build its massive helium core.

The rarity of these events and the lack of a clear connection with stellar metallicity certainly seems to demand a complex and fortuitous set of circumstances. Likewise, the lack of a chemical signature for pair instability in Population II stars implies a highly unlikely set of serendipitous events leading to the detonation of pair-instability supernovae. Both problems would seem to imply that rare stellar merger events – occurring in dense, rare clusters of massive stars such as MGG-11 in M82 – are the root-cause of ultra-rare pair-instability explosions (see Chap. 6). Otherwise, if initial mass or metallicity were the sole drivers of these events we might expect to witness these explosions more frequently, and observe their signature in ancient galactic stars.

Alternatively, better and earlier observations might have shown that SN 2007bi was a rather more mundane, if unusually powerful, core-collapse event, as postulated by Ken'ichi Nomoto and co-workers. If so, all the former questions disappear, and we are left with the simpler question. Do pair-instability explosions happen at all, or are we just doodling math on the back of an envelope? Although there is a tendency to favor the brave in these situations, we shouldn't be too hasty in jumping to conclusions. Pair instability may be an explosion confined to dreams rather than reality. The possibility that conventional core collapse can power fantastically bright supernovae must be addressed. We also need better observations if we are to conclude that this extraordinary event was forged through extraordinary circumstances.

Conclusion

Pair-instability supernovae were formerly the preserve of speculation. It was thought that these explosions, and their pulsational pair-instability cousins, were confined to the universe's earliest history.

However, a smattering of very powerful explosions are hinting that the universe is not quite ready to give up on this part of its history. A number of powerful explosions indicate that very massive stars can produce violent pulses prior to their deaths. These pulses appear to be driven by the formation of matter–antimatter pairs in their cores as the temperatures climb. Another, even smaller group, of explosions appear to be pair-instability events. Whether this is confirmed will be left to further observation.

However, it now appears that the creation of matter–antimatter pairs may play a crucial role in the deaths of the most massive stars – and in the synthesis of elements such as iron, which were formerly thought to be the preserve of Type Ia events. Elucidating the connection between the increasingly broad family of supernovae and the elemental contributions these make will form the basis for some very interesting work in future.

9. The Magnetar Model for Ultra-Luminous Supernovae

Introduction

At the heart of the majority of core-collapse supernova explosions lies a young neutron star. This is formed through the collapse of the 1.44–3 solar mass iron core left behind at the end of silicon fusion. When this body of iron grows sufficiently in mass, it becomes unstable to gravity and implodes, forming a remnant star composed primarily of neutrons.

Zombie stars like this pulse with waves of radio, visible and X-rays in a regular fashion. This is dictated by their rotation and their powerful magnetic fields, which are perhaps one trillion times the strength of Earth's (10^{12} G). Appearing as pulsars, these objects are the conventional end-point for the majority of core-collapse explosions.

Although many supernova remnants appear to lack a pulsing heart, it is generally assumed that the beams of energy, which are manifest as pulsations, are not pointed towards Earth and are, therefore, undetectable. The neutron star-pulsar is still present. It's simply that we can't see it.

Once the supernova shockwave has disrupted the star one normally expects this object to slowly cool and fade from view over the course of a few million years. Initially spinning at tens of times per second, the neutron star's magnetic field slowly brakes it against the interstellar medium, until its rotation grinds to a halt. At some critical point in this slowdown, the magnetic field dwindles, and the beams of radiation fade, causing the pulses to stop.

However, a few neutron stars (perhaps 10 %) are endowed with a magnetic field a hundred trillion or more times the strength of that of Earth's (10^{14}–10^{15} G). These magnetars are a world apart.

D.S. Stevenson, *Extreme Explosions: Supernovae, Hypernovae, Magnetars, and Other Unusual Cosmic Blasts*, Astronomers' Universe,
DOI 10.1007/978-1-4614-8136-2_9, © Springer Science+Business Media, LLC 2014

The magnetic field strength is so intense that it will distort the normally round atomic nucleus into a cigar shape, aligning it with the field of the star. In turn, the field periodically shatters the neutron star's normally rigid iron crust, generating starquakes. During these events the crust releases its pent-up store of energy in a massive wave of gamma rays. These are the so-called soft gamma ray bursts.

Fascinating though these processes are, many astronomers are more interested in the origin of these peculiar neutron stars and, moreover, how these may enhance the luminosity of supernovae. Stan Woosley and others have suggested that these enigmatic magnetars may power the brightest supernovae. Researchers suggest that it may be the birth pangs of magnetars, rather than pair instability collisions or black hole-driven jets, which power many of the most luminous events.

In Chaps. 3 and 8 we looked at a number of very luminous Type IIn supernovae. In the original study on SN 2006tf (Chaps. 3 and 8), a magnetar was considered as a potential driver of the explosion. Moreover, two other supernovae, which were accompanied by X-ray flashes (Chap. 4), bore some unusual features in their light curves. These peculiarities suggested that something odd lay within their expanding supernova debris. Whatever this was, it contributed extra luminosity to each explosion.

So, how can a magnetar power an unusually brilliant supernova?

The Magnetar Model

Let's consider how the magnetar forms. When the core of the supergiant or a Wolf-Rayet star implodes, angular momentum must be conserved. As the rotating Earth-sized iron core collapses, its speed of rotation must increase to match its shrinkage until the object is rotating at a few hundred times per second.

For a few of these nascent neutron stars, the rotation speed appears to cross some sort of threshold, leaving the star endowed with a particularly gruesome magnetic field. At present it really isn't clear what decides this fate. It may simply be dependent on the initial rotation rate of the iron core, the strength of any inherent magnetic field, or the manner in which the neutron star formed

The Magnetar Model for Ultra-Luminous Supernovae 231

from the collapsing stellar debris. Whatever it is, 10 % of neutron stars are born with a combination of these unique conditions that work a little magic, leaving a magnetar as the outcome.

The young neutron star is born hot – perhaps 500 billion degrees. Although it rapidly cools through the release of the ephemeral neutrino, the interior violently convects in an attempt to shed heat through its small surface. It is assumed that rapid internal convection, coupled to the initially rapid rotation, drives the formation of the powerful magnetic field.

Given their presumably initially vigorous rotation, it is their observed *slow* rotation that sets magnetars apart from other young neutron stars when they are picked up in the field. Most magnetars rotate a few times per second. But, if they are born rotating hundreds of times faster, how come we see them, decades to centuries later, with such sluggish rotation?

The key to this conundrum is the effect of their intense magnetic field. A powerful field will rapidly interact with any conducting material around the star, such as the hot plasma from the supernova. This generates powerful electric fields in the hot gas and, consequently, a counteracting strong magnetic field.

You get a similar effect with a transformer – a device used to alter the voltage in electrical equipment. Transformers work as follows. Electrical current flows through a coil, which generates a magnetic field. This is known as the primary coil. This field propagates at the speed of light to a neighboring secondary coil of wire in which a current is then induced. By varying the number of turns of wire we can then adjust the voltages of the electricity from power grid to domestic supply.

To make a transformer work all that is needed is an alternating current – like one in the national grid that reverses its direction every 50th of a second. However transformers, like every other device, are not 100 % efficient. As they work, they heat up. Energy is shed to their surroundings. The cause of this effect is relatively simple in principle. As the current is induced in the secondary coil of wire, it generates its own magnetic field that opposes the field in the first coil. The net result of this conflict is heating within the coils. In a supernova explosion, magnetic fields are present on the order of trillions of times stronger than those found in generators. Consequently, the heating effect is much more severe (Fig. 9.1).

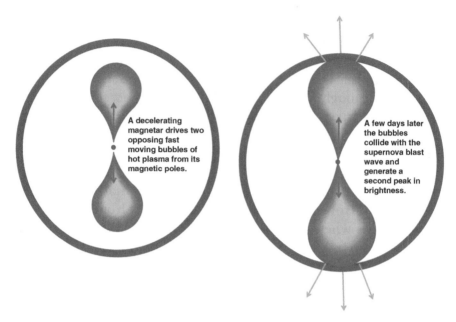

FIG. 9.1 The magnetar model for some ultra-luminous supernovae, and the modestly bright SN 2005bf. The initial rise is caused by the expansion of the supernova shell and radioactive decay. The secondary rise a few hours to days later is caused by the collision of magnetar bubbles with the slower moving, outgoing supernova shock. There was evidence for only one such bubble in the spectrum of SN 2005bf. However, there could have been a second bubble, occluded from our view – or the bubble that was detected was produced by sonic waves that accompanied the collapse of the star's core (Adam Burrow's model – Chap. 2)

Models by Daniel Kasen (University of California, Santa Cruz), Lars Bildsten (Kavli Institute for Theoretical Physics, University of California) and others suggest that over the course of hours to days after the magnetar is formed, the interaction of the magnetic field in the magnetar and that in the surrounding plasma brakes the rotation of the star. All of this liberated energy generates a rapidly expanding bubble of hot plasma. These bubbles sweep outward along the magnetic axes of the nascent magnetar. These propagate faster than the expanding supernova remnant and eventually collide with its outgoing shells of material.

It is this collision that is proposed to generate the additional heating and luminosity powering some of the most luminous events. Most importantly, given the delay from magnetar

The Magnetar Model for Ultra-Luminous Supernovae 233

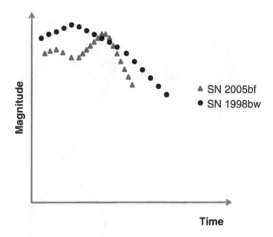

FIG. 9.2 The light curve of SN 2005bf based on Subaru observations, showing a double peak. The first peak occurs around day 20 with the second, higher peak near day 60. Does the expanding shock from a magnetar power the second peak? For comparison the light curve of SN 1998bw is shown

inception to slow down, the magnetar model produces a testable prediction. The supernova will brighten twice. Driven by the collision of the magnetized bubble with the supernova shockwave, the second surge in luminosity will come after the initial peak driven by radioactive decay.

The obvious question is, have we seen any supernovae with such double-peaked light curves? In short, "Yes!" SN 2005bf is perhaps the best characterized of these double-peak events. However, interestingly, many researchers also point to another supernova, SN 2006aj, with its single peak, as evidence for the presence of a magnetar. This may sound contradictory, yet the explanation of this apparent contradiction may illuminate another type of explosion considered in Chap. 4 – the X-ray flash, or XRF (Fig. 9.2).

Let's first try and relate SN 2005bf to the magnetar models of Kasen, Bildsten and Stan Woosley. In these, magnetars with the more modest field strengths of 10^{14} G spin down more gradually than those with the most extreme fields (10^{15} G). The supernovae, accompanying their formation, then have time to peak before the second, magnetically driven blast wave catches up and impinges upon it.

More powerful magnetars will generate their bubble of plasma at a far faster rate, and thus the collision of the two will occur earlier. In turn, this makes the effect of the collision much harder

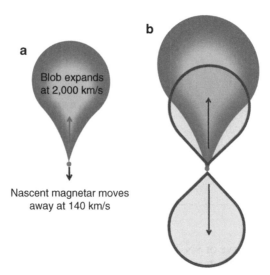

FIG. 9.3 Maeda's best fit model for SN 2005bf. The initial supernova is highly asymmetric, generating a hot bubble of gas that dominates the supernova light curve's first peak at 20 days. The bubble may have been produced by sonic waves that accompanied the collapse of the star's core (Adam Burrow's supernova model – Chap. 2). The magnetar was modeled to have spun once every 10 ms, with a magnetic field strength of 10^{14}–10^{15} G. At 40 days the initially fast-spinning magnetar has shed most of its rotational energy into large secondary bubbles of plasma that reach the supernova debris, energizing them once more. Once these have swept past the debris, swallowing most of it, the light curve resumes its rapid decline

to dissociate from the initial peak of the supernova. The supernovae centered on magnetars with initial field strengths of 10^{14} G are likely to display double peaks.

Was there evidence of this in the light curve and spectrum of SN 2005bf? We already know that the light curve had two peaks: the first at day 20, with the second at around day 40. The latter had an absolute magnitude of −18.3, marginally brighter than the first. Such an unusual feature is rare in the light curve of any supernova, and one which cannot be explained by standard heating models involving the decay of nickel-56 (Fig. 9.3).

The overall light curve suggested that a modest 0.08 solar masses of nickel-56 had been synthesized. Yet, the peak absolute magnitude demanded considerably more than this – around 0.3 solar masses. Thus the presence of a second peak could not be reconciled with the amount of radioactive nickel produced. What about a simple collision? As we have seen in Chaps. 3 and 8, the double peak could be explained if the blast wave had swept outward and then collided with a more distant shell of matter, much as was thought to power events such as SN 2006gy.

However, Kei-ichi Maeda (University of Tokyo) and colleagues analyzed the spectrum of the supernova and identified a fast moving (blue-shifted) bubble of matter directed towards us, but buried *within* the shell of the supernova. This arrangement clearly ran contrary to expectation. We would expect the fast-moving material to have already collided with the supernova's shockwave. In all likelihood, this bubble spawned the second peak in the supernova's light curve when it collided with the outgoing supernova debris. By inference a magnetar had spawned the internal bubble, which had reinvigorated the supernova's light curve. Taken together, this neatly ties theory with observation. The observation of the bubble negated both the pulsational pair-instability shell model and anything involving simple radioactive heating. Simply put, the magnetar model was the most likely explanation for the double-peak supernova.

SN 2006aj and the X-ray Flashes

If SN 2005bf is a good match for the magnetar model's predictions, what about the single-peaked SN 2006aj? How can this be explained using the magnetar model? Finally, are there any other supernovae that might be unusually bright because they host a young magnetar?

Where a supernova creates a very strongly magnetized magnetar (10^{15} G) the magnetar becomes locked to the outgoing shockwave at much earlier times than one where the field is somewhat weaker. Thus these powerful 10^{15} G magnetars decelerate much more rapidly than those with weaker fields. Consequently, the magnetar delivers most of its rotational energy very early on, as the supernova is still brightening under the influence of its expanding photosphere and abundant radioactivity. Extra early luminosity, or the generation of bipolar jets, would be the immediate consequence of the powering down of these magnetars.

SN 2006aj peaked at an absolute magnitude of −19.02 and was associated with GRB 060218. The spectrum of this supernova showed the presence of abundant radioactive nickel (0.21 solar masses) – something frequently associated with GRB-supernovae – but most interestingly a fraction of this material (0.02 solar masses) was directed in broader, likely bipolar, outflows. Moreover spectra implied that a sizable mass of oxygen was directed in jet-like structures.

However, the energy in this GRB was substantially lower than in conventional long events (Chap. 4). Paolo Mazzali (Max-Planck Institute for Astrophysics) and co-workers' analysis of the explosion suggested that this supernova owed its origin to the formation of a magnetar within the heart of a 20 solar mass star. Such a star has roughly half the mass of the type of giant that is expected to give rise to black holes.

Working through the statistics on XRF events, both Paola Mazzali and Alicia Soderberg (Harvard) reached the same conclusion. X-ray flashes (XRFs) may be associated with the birth of all magnetars (Chap. 4). In defense of this assertion – one that cannot be proven by direct observation – Alicia Soderberg noted that the observed rate of XRFs broadly correlates with the rate of formation of magnetars in the Milky Way. Thus it is possible, if not certain, that all XRFs signify the formation of magnetars, while long GRBs are the birth cries of more massive, stellar black holes.

It therefore appears that we can observe the birth of two of the universe's most enigmatic objects through the violent dismemberment of their mothers. Although they may be an order of magnitude or more apart in terms of energy, they share a common underlying process that astronomers are now feeling they are beginning to understand.

Conclusions

The models of Blidsten, Kasen and Woosley, developed only a few years ago, have come a long way in terms of observational confirmation. Some supernovae display the clear impact of magnetar formation and spin-down. Indeed, as Stan Woosley pointed out in a paper in 2009, it is hardly surprising that neutron stars

The Magnetar Model for Ultra-Luminous Supernovae 237

contribute in this fashion to supernova displays. They are an immense store of rotational and magnetic energy, which if tapped at an appropriate time, must produce a supernova-like display in their own right. Although the impact of a magnetar on supernovae such as SN 2006aj is more circumstantial, there appears clearly a role for such an interaction in the case of SN 2005bf. Close scrutiny of other supernova light curves will undoubtedly reveal more of these magnetar-powered, double-hit explosions.

10. The Mysterious SN 2005ap and Luminous Blue Flashes

Introduction

What's blue, distant and very, very brilliant? With a growing litany of supernovae, one might imagine the discovery of another bright supernova in 2005 would simply add another chapter to this expanding tome, one which would soon fall into line with at least one of those already characterized. However, SN 2005ap proved to be a very troublesome supernova indeed – and soon it wasn't alone.

SN 2005ap formed part of a small but expanding nest of supernovae for which there are currently no convincing explanations for their brilliance. At present all we can say is that these enigmas are all fairly distant events with moderate red-shifts, each has a linear light curve resembling the Type II-L SN 2008es, each is very blue in color and hence appears to be energetic, and finally each occupies the pinnacle of present supernova luminosity. This is, broadly, where the boundaries of our understanding lie. At the time of its discovery, the prototype event, SN 2005ap, was very poorly understood.

Although we presently cannot be certain what really powers these events, a number of explanations borrowed from preceding chapters might fit with the relatively sparse facts we currently have.

SN 2005ap

SN 2005ap was something of an mystery from the moment Robert Quimby discovered it on March 3, 2005. The supernova was found using the Texas ROTSE IIIb robotic telescopic at the MacDonald Observatory. Initial analysis of its spectrum suggested that it was a Type II-L event, and one with unusual power. Initial estimates of its luminosity made it the brightest supernova on record, with an absolute magnitude of more than –22. After a brief hiatus, when SN 2006gy was thought to have outshone it, re-evaluation of SN 2005ap's brilliance meant that it regained its crown as the most luminous supernova, with a whopping peak absolute magnitude of –22.7 – roughly twice the luminosity of SN 2006gy.

This unusually blue supernova had an uncertain rise time, but one which was probably relatively short at around 1–3 weeks. Spectra from the explosion showed a peculiar bland continuum with only faint absorption dips, which meant classifying the explosion was difficult. Given that the overall shape of the light curve, with its unusually fast decline, was reminiscent of a Type II-L event – and the weak dips in the continuum could, in principle, be matched to hydrogen – the supernova was then classified as Type II-L by Robert Quimby in 2007. This information is still found on many websites today. It later emerged that this classification was erroneous, but that was hardly surprising given the event's cryptic nature.

SN 2005ap was unusual in other regards as well. Although clearly not a "hypernova" – a supernova associated with a detected GRB or XRF – debris within the blast displayed unusually high velocities, approaching 20,000 km/s, and the temperature of the ejecta was also higher than normal, exceeding 14,000 K.

At the time of discovery luminous Type II-L supernovae, such as SN 1979C, were not without precedent, but this one was a record breaker. Comparison can be made with SN 2008es, which erupted 3 years later. Both shared a similarly bright peak magnitude and fast rate of decline. However, SN 2008es clearly displayed the spectral signature of hydrogen and other elements, and its light was considerably less blue overall. So, without help from hindsight, could the eccentric SN 2005ap, with its high energies and bland spectra, be a unique Type II-L event?

Re-evaluation of the spectra of SN 2005ap brought surprises. Those peculiar bumps in the bland blue continuum weren't hydrogen after all. The explosion was hydrogen (and helium) free. SN 2005ap was, seemingly, a unique form of Type Ic supernova.

Table 10.1 The overall characteristics of each of the enigmatic supernovae

Supernova	Supernova class	Redshift	Peak absolute magnitude
SN 2005ap	Type Ic	0.283	−22.7
SCP 06F6	Type Ic	1.189	−23.5
PTF 09atu	Type Ic	0.501	−22.0
PTF 09cnd	Type Ic	0.258	−22.0
PTF 09cwl (SN 2009jh)	Type Ic	0.349	−22.5
PTF 10cwr (SN 2010gx)	Type Ic	0.230	−21.7

As SN 2005ap was being re-examined another, even more brilliant and mysterious explosion was caught. Kyle Barbary (Supernova Cosmology Project) identified and characterized SCP 06F6. The Supernova Cosmology Project is another supernova project that utilizes a pencil beam search by the Hubble Space Telescope. SCP 06F6 had a strangely symmetrical light curve, with rise and decline times of roughly 60 days.

Despite the difference in the shape of the light curve, SCP 06F6 and SN 2005ap shared the same odd blue continuum and a distribution of spectral bands that were prominent by their absence. Yes, there were a few dips and bumps, but once again, they were weak and difficult to place. Did these two explosions share kinship?

The peculiar shape of the light curve – quite unlike any other supernova of its time – led to a number of unusual explanations for this event. Note the use of the word "event" rather than "supernova," as it wasn't clear when it was discovered if SCP 06F6 was a supernova. Suggestions for this explosion ranged from the collision of an asteroid with a white dwarf somewhere in the galactic halo to an outburst from a distant quasar. More ad hoc explanations such as a Type Ia supernova detonating within the carbon-rich wind of a carbon-star were also mooted. None of them could truly re-create the odd light curve or spectra. So what was SCP 06F6, and what was its relationship – if any – with SN 2005ap? Enter the Palomar Transient Factory (PTF) and the Catalina Real Time Transient Surveys (Table 10.1).

Arrival of the White Knights

Between 2006 and 2010 two automated supernova searches identified several more brilliant blue explosions. The Palomar Transient Factory identified a group of distant events, two of which were independently identified in other searches – SN 2009jh and SN 2010gx. Two of these (PTF 09cwl and PTF 10cwr) were clearly associated with Type Ic supernovae. PTF 10cwr (SN 2010gx) was extensively characterized by Andrea Pastorello (University of Belfast) and provided significant clues that aided in the dissection of the luminous blue puzzle. Pastorello identified numerous emission features in the later spectra of SN 2010gx that identified a clear relationship between these otherwise relatively bland blue events and Type Ic supernovae in general.

Although SN 2010gx initially bore a spectrum that was largely blue and featureless, the supernova eventually shed some of its inhibitions, revealing lines attributed to silicon, iron and calcium. These began to appear after 20 days of expansion as the debris cooled. Spectra were characterized at all epochs by broad P-Cygni profiles, indicative of sustained, high expansion velocities. Whatever was emitting the observed light was moving with a rather extreme rapidity and showed little inclination to stop. Indeed this facet also characterized other PTF discoveries. With the identification of oxygen in these explosions, as well as several other clearly attributable spectral features, the breakthrough with SN 2005ap and SCP 06FC soon followed.

Robert Quimby soon dissected the spectra of the other PTF explosions and revealed that they contained weak emission lines of magnesium attributable to material in the galaxies that hosted these events. With this information to hand, Quimby then determined the red-shifts to each explosion and the rest of the data fell into place. In what could easily be described as a stroke of genius – or at least "a coffee and donuts moment," Quimby realized that the spectra of all of these explosions, SN 2005ap and SCP 06F6 included, could be aligned with the known Type Ic event SN 2010gx (PTF 10cwr). Immediately, the fog lifted, and the enigmatic spectral features of SN 2005ap and SCP 06F6 became clear. These were attributable to oxygen in the supernova blasts. The apparently enigmatic spectral features

The Mysterious SN 2005ap and Luminous Blue Flashes   243

were absorption lines that had been shifted in location around the spectrum by the red-shift of each blast.

Recall that as the universe expands, light from stars and galaxies is stretched by the expansion of space. The further away the star or galaxy is from us the more the light is stretched. Thus, lines of absorption or emission are moved progressively, *en masse*, towards the red end of the spectrum. As long as you know that the light has been shifted, placing spectral fingerprints is straightforward. However, without this knowledge it won't be clear whether you are looking at one spectrum or another.

Once Quimby had aligned the spectra of each explosion, using the magnesium lines as a point of reference, he could then maneuver the spectra as a whole. This allowed the identity of each of the spectral features of the other PTF events, SCP 06F6 *and* SN 2005ap, to be made. In turn the red-shift – and hence the distance to each explosion – could be calculated. This was the long-sought breakthrough.

With the spectra aligned it became clear that these unusual Type Ic events were a distinct class of explosion. Their peculiar continuum; their high expansion velocities, sustained in all but SN 2005ap; and their high temperatures placed them in a unique category of hydrogen- and helium-deficient events. What was also unclear was the source of all that luminosity ($1-2 \times 10^{44}$ J of radiated energy in each case). The broadly symmetrical light curve and high peak luminosity were incompatible with a radioactive power source. For such a high peak, the swift, linear decline, seen in all of these explosions, conflicted with expectations of the exponential decay of cobalt and nickel-56.

However, if a collision was the source of the luminosity, why was there no evidence that the ejecta were slowing down (SN 2005ap, excepted)? The only suitable explanation was that an already fast-moving shell was tailgated by another moving swiftly outward from the center of the explosion. But what was its origin?

Any of the scenarios that involved the decay of copious amounts of radioactive nickel-56, such as an energetic core-collapse or pair-instability event, were ruled out by the shape of the light curve. So, too, was a simple pulsational pair-instability event as was proposed for SN 2006tf and SN 2006gy – unless, and this is an important caveat, these stars had previously undergone outbursts and shed their hydrogen- and helium-rich envelopes. Moreover, the previous pulsational pair-instability explosions showed clear evidence for dramatic decelerations in their blast waves upon encountering the previously shed outer layers.

In each of the likely cases (SN 2006gy, SN 2006tf and possibly SN 2008am) the blast waves moved at velocities less than or equal to 4,000 km/s following the collision. Those of the luminous blue transients moved up to ten times these values, and (with the exception of SN 2005ap) showed little or no inclination to stop. This meant that either the collision imparts so much kinetic energy to the impacted layer that it suffers little loss in its outward momentum when energy is converted to light; or that the impacting inner shell is so massive compared to the outer, tailgated shell that relatively little kinetic energy is transferred in the collision.

So what model fits? There are actually two models that we have already encountered in earlier chapters: the pulsational pair-instability model and the magnetar model. Both can be made to work, explaining these blue explosions – *if* they are suitably tweaked. Let's look at each model in turn (Fig. 10.1).

The PPI Model

If we take the pulsational pair-instability (PPI) model, the stratagem is successful if the star has evolved considerably more than that proposed to account for explosions like SN 2006gy. The progenitor must have already undergone at least two pulses, ejecting first its hydrogen- and then helium-rich layers. The rump carbon-oxygen Wolf-Rayet (WR) star can then undergo additional outbursts if its mass is sufficiently high (greater than 48 solar masses). Such a star ejects a chunk of its carbon-oxygen layer in a penultimate blast while a WR star. A short time (less than a few years) later the core finally gives way, and the blast wave of the supernova impacts the carbon-rich material ejected a short time earlier.

A key point with the PPI model is that the ejected shells become less massive but faster moving with each pulse. This is because each shell has broadly similar amounts of kinetic energy, but each successive shell is less massive than the one before it. With less mass, each shell will, therefore, move faster. This would

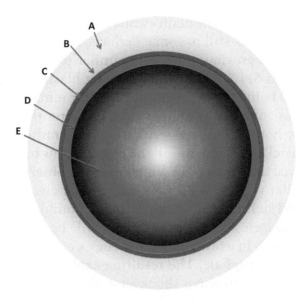

FIG. 10.1 A possible anatomy of the SN 2005ap-like Type Ic supernovae. A highly evolved and very massive star has shed its outer hydrogen- and helium-rich layers. The remnant star is still very massive and is embedded in a carbon-rich wind (A). When the star explodes a shockwave moves rapidly outwards (B), compressing and heating this layer. A primary photosphere accompanies the shockwave (C). Immediately behind this layer is a reverse shock (D). Further in, and hidden from view (E), is the inner photosphere of the supernova. These supernovae may be repeat-offenders – pulsational pair instability events. – except that the inner shell is far denser and more massive than the layer into which it expands

allow the final pulse to release a fast-moving shell that is then impacted by an energetic blast wave or jet produced in the ultimate core collapse. The material will also be deficient in both hydrogen and helium in accordance with observations.

However, the obvious problem with this model is the mass of the shells. As we have already seen, if the blast wave is not to lose impetus upon its collision, it must be substantially more massive than the material into which it collides. Yet, the PPI model predicts a reduction in the mass of the shell with each pulse. The nature of the pulse changes once the carbon core is exposed, with more energy available.

Perhaps, then, SN 2005ap is a foretaste of what may happen to any remaining star at the heart of SN 2006gy. And, as was suggested, the observed "supernovae" – SN 2006gy and SN 2006tf – were actually imposters, and the main events in each case – the collapse of the stellar core – is still to come. Perhaps, and it's a big *perhaps*, SN 2005ap had a previous explosion similar to, and as bright as, that observed as SN 2006gy. This imposter blasted out the outer hydrogen- and helium-rich layers, but the star remained.

In the case of SN 2006gy what remains after the 2006 explosion is a very hot but still massive carbon star, rather than a neutron star or black hole. This moribund object is now busily evolving toward its final decimation, in which case the observed pulse and pursuing stellar wind will be sweeping up and clearing away the hydrogen- and helium-rich material from around the star. Over time this hydrogen and helium-rich gas will be replaced with a carbon-rich vapor. In a few years SN 2006gy's core will finally give up the ghost and implode. The resulting explosion will pulverize the star and blast a massive shockwave through the carbon-rich wind that surrounds it. With a low overall density, the luminosity rises and falls symmetrically, and the hot material produces a relatively bland spectrum. Thus the mighty SN 2006gy would be followed by an even mightier SN 2005ap-like blast.

If this model is true then all of these peculiar Type Ic explosions will eventually reveal the presence of the outer helium and more distant hydrogen layers that were cast off earlier. When the carbon-rich blast wave plows into these more distant shells of debris, much of the shockwave's energy will excite these layers. This is much in the manner that the outgoing shockwave of SN 1987A is now impacting its distant ring of hydrogen-rich gas. Thus, brilliant though SN 2006gy and SN 2006tf were, the magnificence of these eruptions might soon be overshadowed by their terminal events – an SN 2005ap-like supernova. Time will tell.

The Magnetar Model

An alternative involves the formation of a magnetar (Chap. 9). Here a magnetar drives a bubble of superheated plasma outward at high speed. This catches up with and collides with the outgoing supernova shell.

The advantage of this model is that the supernova can occur naturally in a WR star without the need to alter our understanding of the evolution of these objects. In order for this model to work the WR star has to implode with sufficient angular momentum to spin up the resulting magnetar close to the breaking point. It must spin on its axis once every 10 ms and have a magnetic field in the ballpark of 10^{14} G. If these conditions are satisfied the magnetar brakes rapidly, shedding sufficient angular momentum that, in turn, drives a powerful bipolar blast. Within days of the initial supernova, the magnetar bubbles impact the fast-moving supernova shell, driving a second wave of luminosity. These finely tuned conditions may in reality be hard to achieve, but given the paucity of such brilliant blue bursts, it is possible that such a rare combination of circumstances could produce these odd events.

The Buried GRB Model

Another suggestion for such blue supernovae is the dissipation of GRB jets within the hydrogen envelope of a supergiant star. The brightest Type II-L supernovae, such as SN 1979C or SN 2008es, may be produced this way (Chap. 8). With an initially high mass, the remnant produced by the death of one of these massive stars would likely be a black hole, its birth accompanied by the jets and gamma rays of a GRB.

If this GRB is thoroughly obscured by very massive shells of expanding, absorbent material we may not see it as a GRB, but rather as an energetic supernova. If the mass of the black hole scales with the star – something that is not clear – then a large stellar mass black hole may be the result. If the mass scales with rotation, and hence has a larger impact on angular momentum, then the power of the jets launched by the young black hole may also increase in longevity, power or both. The physics of these young objects is not yet clear. However, the point is that there are ample opportunities to form a violent explosion that have odd spectrum, are hot and have high kinetic energy. The physics needs investigating.

However, contrary to observations, it is expected that the buried central engine would betray its presence through the formation of broad wings around certain spectral emission features.

Crucially these wings are absent in the spectra of the PTF and later Pan-STARRS events (below). Their absence may rule out a collapsar model for these explosions. Certainly the existence of dense shells of carbon-rich matter around these explosions seems hard to explain in the context of a collapsar model. However, some sort of hybrid model may work. PPI events generate the shells, while a later core collapse generates a heavily obscured GRB.

Are These Blue Explosions Pair-Instability Events?

Almost certainly a bona fide pair instability explosion, suggested for SN 2007bi, is ruled out. Such an explosion, one resulting from a star with, perhaps, 200 times the mass of the Sun, would produce too broad a light curve to match those curves observed.

Pan-STARRS Events

In 2010, the automated supernova search Pan-STARRS turned up two further examples of these enigmatic blue explosions: PS1-10ky and PS1-10awr. Like the earlier sample these occurred at modest red-shifts of 0.9 and radiated energies of 10^{44} J, making for a very consistent set of supernovae. Each had a luminosity of –22.5, only marginally less than SN 2005ap, and displayed ejecta moving at velocities of 12,000–18,000 km/s – again very consistent with the earlier PTF and Catalina discoveries.

During the period of observation neither explosion showed any indication of deceleration. Laura Chomiuk (Michigan State University) and colleagues suggested that these Type Ic events had ejected a massive shell of carbon-rich material prior to the supernova. The resulting light curves were explainable if the supernova shockwave was breaking through this shell and out into interstellar space. This again suggested that the shell was ejected prior to the supernova – or was representative of magnetar bubbles breaking through the outgoing carbon-rich supernova shell.

Can We Tie Up All the Loose Ends?

So, which model works best at explaining all of the observations? The magnetar bubble should produce peak luminosity as the shockwave begins to boil or blast its way through the supernova shell, while the PPI model suggests that the shell lies at considerable distance from the supernova and is then hit sometime after the star blows up. The precise details of the resulting light curve depend on whether the light from the photosphere of the supernova infiltrates the shell before the shockwave has fully penetrated and heated it. However, in principle these models can be distinguished.

SN 2010gx (PTF 10cwr) provides some clues. This event bore all of the already well described features of luminous blue transients. However, at later times, the features of the blue transients gave way to the more normal characteristics of a Type Ic explosion. Andrea Pastorello suggested that the initial blue transient phenotype was produced by the outer shell of material, while the later Type Ic features arose as the optically thick (opaque) shell was penetrated and cleared by the supernova's shockwave.

These crucial observations are perhaps best explained by the ejection of a shell followed a short time later by the implosion and detonation of the remaining core, rather than a supernova followed by a magnetar bubble. The latter model is hard to reconcile with observations of an underlying Type Ic explosion and the apparent dissipation of the shell with time, while the PPI model naturally accommodates these observations. However, the consistency in the radiated energies of each of these events might best be explained by a finely tuned magnetar. What are needed are further observations. Pulsational pair-instability events should reveal themselves through multiple explosions prior to the final blue transient and the presence of a final Type Ic supernova.

Given the predicted luminosities of such PPI events each should be readily detectable ahead of any final blue transient. Conversely, the magnetar model predicts a single supernova explosion followed by a secondary brightening or a change in the motion of the debris as the bubbles burn through the outgoing supernova shell. Again, spectra should provide evidence for these magnetically driven explosions.

One of the more interesting features of these supernovae is their relatively consistent, moderate red-shift. Is it coincidental that each occurs at such red-shifts, or is it simply some sort of observational bias? Why don't we observe any more of these explosions at higher red-shifts? These explosions are certainly bright enough to be detectable from further afield.

If there aren't any reasonable selection effects, are there any astrophysical reasons that would favor the destruction of a star in this peculiar manner at a red-shift around 1.0? This might seem unlikely, but perhaps the universe needs time to build the sort of single (or binary) star system that can generate this sort of explosion. Too few of these events have thus far been observed to really be sure whether it is serendipity or hard-line physics that plays the winning cards in the observed red-shift distribution of these explosions.

Finally, are there any suitably massive stars in the modern universe that might produce the kind of multiple PPI eruptions likely to generate these peculiar blue explosions? Well, this depends on our very imprecise understanding of mass loss. In principle the recently discovered hyper-star R136a in the Tarantula Nebula might fit the bill. Based on its estimated luminosity, this object may hold 260–310 Sun's worth of mass. However, this estimate is fraught with uncertainty, as the apparently single star may in fact be a dense massive cluster or pair of still substantial but less massive stars. Even if the estimate of its mass holds up, uncertainty then surrounds how much of this bulk the star can retain through its life. Calculations by Ken'ichi Nomoto and others suggest such a star could lose enough mass to end its life as a conventional core-collapse supernova – albeit one likely to produce a black hole and potentially an accompanying GRB. Once again, a better understanding of mass loss wouldn't go amiss.

Conclusions

In the last 10 years automated supernova searches have moved beyond simply locating more of the same. Now, they are turning up many interesting needles in the many galactic haystacks that populate our visible universe.

However, it must be made absolutely clear just how rare these super-luminous explosions really are. A broad sweep at the numbers provides a compelling picture of sparrows swamping sea eagles. If we set a ballpark figure of 100,000 core-collapse supernovae per cubic gigaparsec per year, we also get around 30,000 thermonuclear Type Ia events. This is still a large number of events. Continuing with our bird analogy, sparrows outnumber the blackbirds roughly 3:1.

Moving into the territory of rarer birds, the Type IIn SN 2006gy-like events clock in at roughly 150 per annum per cubic gigaparsec – or 0.5 % the frequency of Type Ia explosions. Yes, these are rare in comparison to Type Ia events, but the super-luminous SN 2005ap-like events are rarer still. These eruptions account for perhaps one fifth the SN 2006gy-like explosions. This figure is less than one in 4,000 supernovae, per cubic gigaparsec. A needle in a haystack would be far easier to spot.

Yet the growth in computing power, coupled to an army of amateur and professional astronomers plugged into their automated 'scopes, now allows far greater pools of data to be scrutinized. These are exciting times in astronomy. Old paradigms are falling, not with creaks, groans and whimpers but with the impact of bulldozers and dynamite, like so many anachronistic Berlin walls.

However, as the shock of the new emerges, we are increasingly bereft of the answers to explain them. Paradoxically many of the most recent discoveries take us back decades to models once thought unrealistic or inapplicable in our modern universe. Never more so than with these brilliant, blue supernovae. There is a feeling with these blasts that nothing quite fits the bill. These seemingly unique events are challenging our models and the limits of how bright supernovae can be. There is an element of forcing round pegs in, at best, hexagonal holes. Can we adapt older models to fit these observations or do we require a new set of models? Whatever the outcome, the next few years will prove to be very exciting as more and more peculiar events are uncovered as formerly rare and peculiar explosions become more commonplace. With luck, the latter trend will elucidate the mechanism or mechanisms underlying these currently outlandish events.

Part III
Thermonuclear Supernovae

11. Hypotheses and an Oxymoron

Introduction

Until the discovery of the bright and peculiar core-collapse supernovae, described in earlier chapters, Type Ia supernovae were the brightest known explosions in the universe. Astronomers had a pretty good idea about where these explosions arose. However, establishing the precise mechanism by which the explosions occurred, or the type of system that gave rise to them, was proving to be more elusive.

At present there are a number of facts astronomers believe to be true plus a whole she-bang of other issues astronomers wrangle over to varying extents. All would agree on the type of object that gives rise to these explosions – a carbon-oxygen cored white dwarf – but not necessarily the road down which these objects travel to their destruction. Recent studies of these explosions, as well as a more thorough dissection of the population of these explosions, are finally shedding light on many of the more complex issues afflicting our knowledge of these events. They are also pointing their way toward a new understanding of how the central players in this drama can explode and generate different kinds of blasts.

What then do astronomers know, think they know, or simply infer? And what does the raft of new Type Ia-like supernovae tell astronomers about the systems and circumstances that produce them?

What Astronomers Know

Type Ia supernovae are produced when carbon-oxygen white dwarf stars explode. The outcome is a rapidly expanding ball of material, rich in radioactive nickel-56. Many texts refer to these supernovae as "standard candles," implying that they are uniformly bright.

This is untrue. However, there is an explicit relationship between the peak absolute magnitude of the supernova and the time it subsequently takes the light to dim. The brighter it is the longer the time the explosion takes to dim. This is a consequence of the power source: nickel-56. The brighter the supernova, the more nickel-56 will be present, and the longer it will take to decay – hence the more elongated light curve. This relationship holds true for all but the brightest, rarest and possibly super-Chandrasekhar explosions. These explosions are too bright for the width of their light curves, and the reason for this discrepancy is still unknown. Thus, in general, Type Ia supernovae can be used to determine distance to galaxies far out in space. Significantly, they were the key that opened the box on our accelerating universe. Consequently, there has been a focus on finding out more about the mechanics of these events, in part to ensure that we are not somehow being hoodwinked into believing in an accelerating universe that proves to be a mirage created by our misunderstanding.

In the category of what astronomers think they know is the assumption that Type Ia supernovae occur when the mass of the carbon-oxygen white dwarf exceeds a particular mass limit (1.382 solar masses). At this point the carbon becomes so compressed that nuclear reactions begin. The evidence for this is theoretical but generally regarded as a pretty solid "fact." Initially proposed by Fred Hoyle, following his seminal work with William Alfred Fowler, Margaret Burbidge, and Geoffrey Burbidge on nucleosynthesis, this presumption has stood the test of time. Destruction of carbon-oxygen white dwarf stars produces Type Ia supernovae.

Moreover, Type Ia supernovae appear to erupt in binary systems, but with considerable variation in the properties of the binary partner. Type Ia events are equally found in both old and young stellar populations. This implies that the stars giving rise to these events are not massive. Such hefty stars simply don't live long enough to be found in old systems. Most importantly, this fact ties the explosions to some kind of event that occurs on the white dwarf star itself.

The energy released in each explosion is in the range $1–2 \times 10^{44}$ J, and on the whole this is fairly consistent. Finally, the contribution Type Ia supernovae make in terms of universal nucleosynthesis is broadly understood in these explosions. Type Ia supernovae produce the bulk of the iron in the universe. They are prolific producers of iron-peak elements; with a high level of consistency, each explosion produces 0.6 solar masses of nickel-56 – hence the similarity in peak luminosity. The weird and wonderful exceptions to this rule are discussed in subsequent chapters, but the bulk of Type Ia supernovae follow this pattern.

Finally, as the name suggests, Type Ia supernovae lack hydrogen and helium in their spectra. There is evidence in some explosions for carbon, but the majority of the products that dominate the spectra are silicon, iron, sulfur and intermediate mass elements such as calcium, titanium and phosphorous.

What Astronomers Think They Know

One thing in the camp of what astronomers think they know is that these supernovae arise when a carbon-oxygen white dwarf reaches (or at least approaches) the Chandrasekhar limit, through the accretion of matter from a companion – the nature of which is largely unknown. The explosion itself must be some combination of deflagration and detonation. A deflagration is a combustion reaction where the material expands at subsonic speeds as it burns. Deflagrations are driven by the transfer of heat by conduction and radiation from the burning front of the material into the surrounding unburned fuel. A detonation is a supersonic explosion where material is violently compressed and heated to the point of ignition. This happens through the progression of a shockwave through the fuel rather than by the direct transfer of heat. To put this in an Earthly perspective the former is characteristic of gas engines, while the latter is found in diesel engines.

The problem with Type Ia explosions is knowing how to tweak the conditions so that the correct combination of detonation and deflagration occurs. This is essential; otherwise the elements synthesized in the explosions do not match the observed abundances. Above and beyond this, a few rare explosions appear to be the result of pure deflagrations, while others show unusual chemical abundances that are hard to explain by conventional Type Ia mechanisms. Therefore, how do we explode the star in just the right way to produce the supernovae we observe? Clearly, the manner in which the star dies must relate to the mode in which the white dwarf star accretes matter, heats up and then, finally, succumbs to the internal stresses caused by ignition of its carbon fuel.

What Astronomers Don't Know

What isn't clear is why there is variation between these explosions. If they derived from the same underlying mechanism, wouldn't they be intrinsically similar? Yet, the amount of nickel-56 synthesized clearly varies. These supernovae, although similar, are not clones. There are certainly assumptions and beliefs, but at present the underlying cause of the variability is not known for sure.

Furthermore, there exists the thorny problem of actually blowing up a dense, planet-sized star. What actually happens to drive an explosion powerful enough to lift degeneracy within the dwarf and blast its entire contents into interstellar space?

We also don't know what sorts of binary systems give rise to Type Ia supernovae. Just how do you set up an explosion, free of hydrogen and helium, if the white dwarf is accreting this sort of material from a companion? If the explosion follows the merger of two white dwarf stars, why is there so little carbon present in the supernovae debris? Imagine blowing up a building and finding there wasn't a shred of concrete in the wreckage! Finally, how do the peculiar Type Ia events (Chaps. 12 and 13) fit into the overall picture? What makes them so different?

Mechanisms and Scenarios

The mechanisms of Type Ia supernovae can be broken down into two, discrete, general phases. In the first phase the white dwarf is accreting matter and slowly but surely edging closer to ignition.

The white dwarf almost certainly is made up of carbon and oxygen. The rarer but more massive white dwarf stars, made of oxygen, neon and magnesium, probably never become hot enough to ignite their fuel before collapsing to form neutron stars. This so-called accretion-induced collapse (A. I. C.) is expected to produce neutrons stars with relatively low masses (1.2–1.3 solar masses) and may account for a proportion of supernovae that fall either into Type II or Type Ib groups (Chap. 7). The precise type of event will depend upon the material that is being accreted. Regardless, the predicted chemical signature of these sorts of explosions does not match the observed abundances in Type Ia events. A. I. C. supernovae probably account for 4 % of core-collapse events (Chap. 7), a value that is dwarfed by the relative abundance of Type Ia supernovae.

In general all Type Ia models assume that the white dwarf star, destined for supernova, is part of a binary system, one where both stars orbit one another so closely that material can be transferred onto the surface of the white dwarf. This allows the white dwarf to grow in mass and reach the Chandrasekhar mass limit.

So, what sorts of companion stars could feed the white dwarf with material that adds to their mass? Again, there is a bewildering variety of possible scenarios, all of which have advantages and disadvantages. That said, analysis of some recent, unusual supernovae, is shedding light on how particular classes of companions may give rise to unique subgroups of Type Ia events.

How do you build up the mass of the white dwarf? There are probably two roads leading to Rome. In one, the companion is another white dwarf, comprising helium, or carbon and oxygen. This is the double-degenerate scenario. In the other, a more conventional star, like the Sun, is in a tight orbit with the companion. In a final variant of this scenario, the companion is a red giant or sub-giant star in a more distant orbit. In each case, the companion star loses matter to their small, dense and more evolved white dwarf star. This is the single-degenerate scenario.

The Single-Degenerate Scenario

In the single-degenerate (SD) route, only the moribund white dwarf is compact and degenerate. In this state the atoms are packed so tightly that the electrons are shunted into specific energy levels and provide pressure that supports the white dwarf star. Although the companion may have a degenerate core, the "donated" (or more accurately stolen) material from the companion does not. In most scenarios the material accreted is hydrogen-rich, but conceivably the material could be helium, siphoned off a more evolved helium-burning star that had already lost its outer hydrogen-rich layers. Either way, this material is assimilated, presumably via some form of disc, onto the surface of the white dwarf. Here hydrogen is first burned to make helium and then fused once more to make carbon, adding to the mass of the degenerate core. However, this scenario has its problems. Most significantly is the issue of helium fusion. To understand this we need to look at how the white dwarf would grow in mass in the single-degenerate scenario.

There is little problem with hydrogen fusion. It can passively add to the mass of the star if the rate of fusion of hydrogen occurs at more than one hundred thousandth of a solar mass per annum. The white dwarf will then appear as a super-soft X-ray source with a surface temperature of 100–200,000 K. In these stars steady stellar winds balance the inflow of hydrogen with its consumption in fusion reactions at the base of the layer. If the rate is much less than this, the white dwarf will undergo unstable, explosive burning and suffer periodic novae explosions. These are expected to erode the mass of the white dwarf, making it unlikely that such white dwarfs could ever grow massive enough to explode as supernovae. If the rate of hydrogen capture is much higher than one hundred thousandths of a solar mass per year, the white dwarf will acquire so much hydrogen that it will either expand back to become a red giant or become consumed in a common envelope, with its withering companion. As such, the single-degenerate model needs extensive fine tuning to make it work well, even at this stage.

Yet, this isn't the only problem. Despite the potential for stable hydrogen burning, helium fusion is another business altogether. The underlying helium layer is likely to become degenerate long before it can ignite. Once this layer eventually becomes hot and dense enough to fuse, the resulting helium shell flash should cause the white dwarf to rapidly balloon up to red giant proportions, much as is seen in born-again giants such as Sakurai's object (Chap. 13). Not only should this prevent further accretion, until the hydrogen and helium layers are either consumed or driven off, but potentially the mass of the white dwarf could be reduced through mass lost from its expanded outer layers during this explosive transformation. Such a helium flash is a major obstacle to the success of the single-degenerate scenario.

Mass loss through these mechanisms has other consequences for the binary star system. As stellar winds whittle away at the born-again red giant (or the super-soft X-ray source), the two stars should drift further apart in response to the overall lowering in the mass of the star system. Hence the reduced gravitational pull of the stars upon one another will increase the distance between the two stars, and this in turn will lower the rate at which mass can be accreted by the white dwarf. The single-degenerate system needs a lot of fine-tuning that, in principle, could either prevent it contributing at all, or limit its involvement in the formation of Type Ia events. Certainly, these problems are giving astronomers headaches.

Moreover, there is an observational problem with the single-degenerate scenario. Elliptical galaxies, with predominantly old stellar populations, tend to host lower luminosity events, each with a narrower (low-stretch) light curve, while young, star-forming systems produce brighter, broader explosions. This suggests that there is a difference in the yield of radioactive nickel-56 in each environment. Why are they different? Is this down to the sorts of progenitor star systems?

Still, it's not all doom and gloom for this model. In a few instances the common-envelope scenario may yield a Type Ia event directly. During the common-envelope phase the white dwarf is consumed within the envelope of its donor. Frictional forces, coupled to a complex interplay of angular momentum between the white dwarf, the donor star core and its envelope, cause the white dwarf to spiral inward. Conservation of angular momentum causes the outward dispersion of the hydrogen-rich envelope of the other star as the white dwarf moves inward. The leftovers from this process consist of the slightly more massive white dwarf and the dense core of the other star. This remnant core appears as a small, hot sub-dwarf star or another white dwarf, depending on when in the life of the companion the common-envelope phase occurred.

In general, if the white dwarf accretes enough material in this phase it may explode, dispersing not only its own mass but some of the companion as well. Hypothetically, this additional material, with its very distinctive chemical composition, may become apparent within the expanding debris of the supernova.

Despite one or two possible events linked to common envelope phrases there is little evidence for a common envelope mechanism generating the vast majority of Type Ia events.

Moreover, the remnant companion star should survive the blast – albeit rather battered, heated and, in the case of a former red giant companion, stripped state. However, the evidence for a surviving companion in some Type Ia supernovae is, at best, equivocal. For example, it was claimed that a Sun-like star was observed departing the scene of the Tycho supernova remnant. This claim was contentious and is not currently – universally – accepted. Moreover, the SD scenario predicts that the dying system will be surrounded by a hot wind of hydrogen- and helium-rich debris emitted by the growing white dwarf prior to its demise. This has not been observed in the spectra of any Type Ia event. Indeed the problem with the SD route as a whole is the requirement that the hydrogen and helium that presumably is accreted by the white dwarf is lost (or destroyed). None should be visible in the subsequent spectra of the supernova. This form of extreme unction seems difficult to achieve, and much more work will be needed to convince many astronomers that the SD route can really work for most Type Ia events.

More Problems

If the companion star was a hydrogen (or helium)-rich star with a non-degenerate stellar structure then in most instances the star system giving rise to the supernova will be embedded in the stellar wind of the normal companion and the wind blowing accreting material from the white dwarf. As the supernova shockwave blasts outwards it should, in theory, plow this material into a dense layer that reveals its presence in the spectra of the explosion, and this in turn should generate a spike in ultraviolet emission early in the supernova. This is akin to shock break-out in a Type II supernova – something that has been observed. However, in almost every Type Ia supernova this is not seen. There is no observed excess in ultraviolet emission at early stages.

Furthermore, a typical diagnostic for Type II supernovae is radio and X-ray emission from material swept up by the blast wave.

Radio emission is also lacking in all Type Ia supernovae, and X-ray emission is restricted to a precious few. In some cases the Type II explosions show narrow and intermediate width emission lines produced by flash ionized material that is then swept up by the blast (Chaps. 3 and 8). However, evidence for this is notable for its absence in Type Ia explosions. A few Type Ia supernovae show evidence for slow-moving material ahead of the shockwave in the form of sodium lines indicative of circumstellar gas. However, this spectral evidence is equivocal and could result from material not associated with the star system, and must, therefore, be treated with caution. So at the moment evidence linking Type Ia supernovae to single-degenerate progenitor systems is weak.

A Precious Few

Despite these many problems there exists a smattering of exceptions, explosions that do show evidence for hydrogen – or at least some type of circumstellar material. A few Type Ia supernovae, such as SN 1991T or SN1999aa, are unusually bright a month or so after the explosion, giving rise to a plateau in their emission. This can be interpreted as evidence of circumstellar gas interacting with the outgoing shockwave. A precious few other Type Ia supernovae provide less equivocal evidence in support of the single-degenerate scenario.

SN 2005ke was a step up this ladder. This supernova showed profuse X-ray emission from days 8 through 120 after detonation. The only plausible explanation for this observation was that the shockwave was penetrating and sweeping up hydrogen-rich material in its path. Stefan Immler (University of Maryland) suggested that this was evidence for the supernova interacting with material shed at a rate characteristic of a red giant or asymptotic giant branch (AGB) star. By implication, the white dwarf had reached the Chandrasekhar mass by accreting material from this wind or directly through an accretion disc from the companion's envelope.

More convincingly still was SN 2002ic. The spectra of the supernova evolved from a fairly normal Type Ia explosion to that of a Type IIn event. Mario Hamuy (University of Chile) and co-workers did an analysis of the supernova which suggested that the blast wave interacted with hydrogen-rich material from a 3 to 7 solar mass AGB star. Again, the suggestion was that the white dwarf had entered a common-envelope phase with the red giant. This phase had ended shortly before the white dwarf detonated. The dispersing red giant envelope was hit by the supernova shockwave a few days after the explosion, generating the narrow and broader emission lines. Strong narrow and broader emission lines tallied with other Type IIn supernovae, but the remainder of the spectra was more typical of an SN 1991T/1999aa-like Type Ia supernovae.

Of note here is the suggestion that the controversial supernova SN 1997cy, described briefly in Chap. 4, was in fact a Type Ia event buried within a hydrogen-rich shell or disc of material. At the time, it was tentatively suggested that this was a notorious "hypernova" with an unusually energetic blast. However, reanalysis suggested that the total energy of this supernova was probably not much, if at all, greater than a conventional Type II supernova. Indeed the similarities in the spectra of both this Type IIn supernova and SN 2002ic led Mario Hamuy to conclude that SN 1997cy may in fact have been a thermonuclear explosion, a direct counterpart to SN 2002ic. If true SN 1997cy occurred within a dense, hydrogen-rich shell of material. The resulting Type IIn light curve was actually a composite of an inherent Type Ia supernova buried within the hydrogen-rich debris of its former envelope or that of its companion.

In theory two possible models could account for the relatively massive hydrogen-rich envelope around SN 2002ic – and perhaps SN 1997cy. The first is a conventional Type Ia supernova that occurred within the dense envelope of an AGB star (or simply detonating close to this star); the second is a theoretical Type 1.5 supernova, discussed in Chap. 7.

To recap, a Type 1.5 supernova is an unusual and entirely theoretical kind of event that might occur in some intermediate mass stars. These explosions were predicted to happen if the star's carbon core reached the Chandrasekhar limit while degenerate and in retention of a deep hydrogen-rich layer. Addition of further mass to the core caused it to undergo collapse, ignition and detonation. Given that the Type 1.5 scenario is believed to be restricted to metal-free, or metal-poor, stars, this requirement probably rules it out. This prerequisite is needed if the star's core is to become degenerate before its entire hydrogen-rich envelope is lost through stellar winds.

Fine-tuning stellar evolution to fit our expectations may seem problematic at best. However, detonating a white dwarf while it is within the hydrogen-rich envelope of its companion may not be as farfetched as it first appears. There are two possibilities. If the white dwarf was initially very massive it might, simply, have reached its tipping point at the stage where it became absorbed into a common envelope. However, more convincingly, perhaps the white dwarf spiraled in through the hydrogen-rich layer until it met and merged with the carbon-oxygen core of the AGB star. In this case it is a double-degenerate event rather than a single-degenerate one. An apparently simple explosion may in fact have a more complex origin.

Finally, is there any evidence for a preference between the single-degenerate or double-degenerate routes? Bradley E. Schaefer and Ashley Pagnotta (both at Louisiana State University) examined the Type Ia supernova SNR 0509–67.5 in the Large Magellanic Cloud. They found no evidence for a residual companion star. Certainly any companion brighter than main sequence, spectral class K, would have been detected, and it wasn't. Their investigation eliminated bright main sequence, sub-giant, red giant or helium star companions in the destruction of this particular white dwarf. With high probability the double-degenerate route appears likely for this particular supernova. A similar conclusion was reached for the powerful SN 1006 in our galaxy.

It is clear that several peculiar Type Ia events appear to be linked to double-degenerate scenarios. Although the single-degenerate route may result in some Type Ia supernovae, it does not appear to be the principle route.

Double-Degenerate Models

In the double-degenerate (DD) route the companion is another white dwarf. Given that these two stars are small, dense and recalcitrant to mass loss, they must lie close together if the more massive of the pair is to get even the slightest chance of stealing material from the other. This situation can arise if the two stars were brought close together during a common-envelope phase.

As the white dwarf plunged through the envelope of the other star, it would, on the course of a few thousand years, move into a very snug orbit around the core of the former red giant companion. With the hydrogen-rich envelope now gone, the two white dwarfs could circle one another before succumbing to a peculiar property of space-time.

Einstein's theory of general relativity suggests that space is warped by matter, something that has been beautifully confirmed by a number of observations of binary neutron stars. If two particularly small but massive objects are in orbit around one another, gravitational waves are emitted as they orbit, and these cause the two objects to spiral even more closely together. Millisecond pulsar binaries have proven to be satisfactory proof of this concept, and it is expected to work efficiently for close white dwarf binaries as well. Over tens or hundreds of millions of years the two white dwarfs will edge ever closer together, until the less massive of the two begins to lose material to the more massive but smaller white dwarf star.

Material from the less massive white dwarf will settle into a thick accretion disc, and given the small size of both stars, the shredded white dwarf will be completely consumed in very little time at all. Accreted helium can burn with varying degrees of explosiveness, but potentially add mass to the white dwarf. Accreted carbon and oxygen will burn once the mass of the white dwarf exceeds 1.37 times the mass of the Sun.

However, this is not as clear-cut as it as first sounds. In Chap. 7 we saw that in the most highly evolved red giant stars, those massive enough to burn carbon, this element ignites once the core exceeds approximately 1 solar mass. The carbon must remain relatively cool, collecting passively until the core approaches the Chandrasekhar limit. Otherwise nuclear fusion will transform the white dwarf into an oxygen-neon-magnesium (ONe) variety. An ONe white dwarf would, in turn, undergo accretion-induced collapse. The outcome of this is a neutron star, not an expanding field of debris. The chemical signature of an accretion-induced collapse is undoubtedly very different from that of an observed Type Ia supernova. Therefore, however, as the carbon is accreted the material must remain cool enough so that it doesn't ignite prematurely. Alternatively, carbon must accrete so quickly that the critical mass is reached before it grows hot enough to ignite.

Is there any evidence for the double-degenerate model? In genetics analysis of populations often provides additional information that constrains theories. Similarly, astronomers can restrict the relative contribution of the SD and DD routes to the Type Ia supernova rate. In general the DD merger model matches the populations of the progenitors and the supernovae better than the single-degenerate (SD) route. But as Dan Maoz (Tel Aviv University) and Filippo Mannucci (Osservatorio Astrofisico di Arcetri, Italy) noted in a recent publication, this may require the modification of the types of double-degenerate progenitors that give rise to Type Ia supernovae. Moreover, Mannucci noted that there was a poor match between the numbers, masses and rates of accretion of the observed white dwarf-red giant binary systems to the Type Ia supernova rate. However, Maoz and Mannucci left the door open for some involvement of these systems in a proportion of Type Ia supernovae. Their article also made for an excellent, amusing and very informative comparison between population studies in humans and stars, highlighting the pitfalls of overly simplistic interpretations!

Although there exists two potential roads leading to Rome, there appears to be a growing consensus that the majority of Type Ia supernovae are produced through the double-degenerate scenario. Within this camp there are a variety of different possibilities as to the nature of the two stars that give rise to the explosion. At present it is not clear which types or masses of white dwarf generate which types of Type Ia event. A lot of work remains to be done.

The Explosion

Although astronomers would place money on a Type Ia supernova coming from the destruction of a carbon-oxygen white dwarf, they probably wouldn't stake a claim on the exact mechanism of its decimation. Initial thoughts turned to the idea that carbon burning might start passively once the mass exceeded 1.3682 solar masses.

Carbon would simmer through a curious process referred to as cryonuclear fusion. Cryonuclear fusion broadly translates as *cold fusion*, and it refers to the ability of nuclear fuels to react if they are suitably compressed – even if they are colder than normally assumed would be required to initiate the reactions. At slightly lower and non-degenerate conditions another series of cold (or cool fusion) reactions called pycnonuclear reactions can occur. These are important early reactions in novae and they may be important in some unusual Type Ia-like supernovae.

The extremely dense interior of a white dwarf will allow cryonuclear reactions to begin once the density rises sufficiently high for carbon nuclei to begin to fuse, irrespective of their temperature. In early models it was presumed that these sorts of nuclear reactions would gently accelerate over hundreds or thousands of years until the temperature became hot enough for full-fledged ignition. At this point a deflagration would ensue, with carbon burning vigorously in the stellar core. Then came the problem. The spectra of Type Ia events demanded that part of the explosion be a detonation, and the precise point at which the explosion transitioned from deflagration to detonation was unclear. It was also unclear why such a transition would happen in the first place.

Over the years, various models have come and gone. The first arrived in 1969, a one-dimensional model by David Arnett (University of Arizona) favored a pure detonation. Later models by Ken'ichi Nomoto (University of Tokyo) favored a rapid deflagration. During the 1990s a hybrid model emerged, but there was little consensus on how this mechanism actually came about.

In other models, the transformation from deflagration to detonation involved helium fusion. Helium would ignite in a dense, and nearly degenerate, layer on top of the carbon-oxygen core. In these conditions, the pycnonuclear reactions could become explosive, and a consequent helium detonation might just compress and heat the carbon core sufficiently to trigger a further detonation inside. However, this model failed to produce the expected abundances of nickel-56, iron-group elements and intermediate mass elements (Fig. 11.1). That said, in recent years this model has had something of a reprieve. A helium-shell detonation could well explain some other types of supernovae, those related to Type Ia events (Chap. 13).

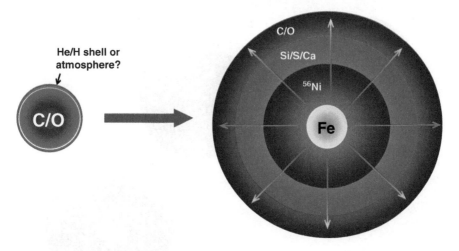

FIG. 11.1 Anatomy of a simplified Type Ia supernova. Once the Chandrasekhar limit is breached the interior of the carbon-oxygen white dwarf incinerates, forming an expanding cloud of iron-rich debris. Further from the center of the white dwarf, where conditions are less extreme, elements of lower mass are produced by the fireball

Prior to the year 2005 no explosion model worked particularly well for Type Ia events. The limiting factor in improving our understanding was computing power. To fully comprehend these explosions they needed to be sensitively modeled in three dimensions – a feat requiring prodigious computation. The potential breakthrough came when the white dwarf was considered in all its complexity. Three-dimensional modeling of a deflagrating 1.3682 solar mass carbon-oxygen white dwarf suggested that ignition would probably begin slightly off center. The divergence from perfect radial symmetry didn't have to be much, but the result was spectacular. The outcome was an enormous, planetary-scale, nuclear mushroom cloud that expanded to consume the entire surface of the star. This burning flow was the key to the success of the "gravitationally confined detonation" (GCD) model developed at the Flash Center (DOE/NNSA ASC Alliance Center for Astrophysical Thermonuclear Flashes at the University of Chicago).

In the GCD model the white dwarf begins to burn through a deflagration as before. However, with gravity acting on the outburst, the two-billion-degree fireball rises rapidly to the surface of the white dwarf. Confined by the intense gravitational field,

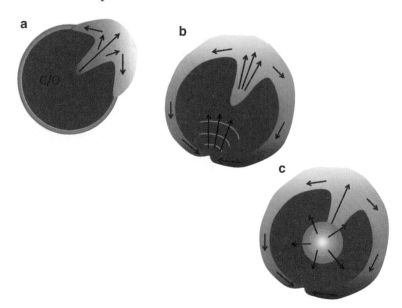

FIG. 11.2 Toward a more realistic Type Ia supernova? In (a) the carbon-rich interior has ignited, and within 1 s a 2–3 billion Kelvin fireball has risen to the surface of the white dwarf and is beginning to spread outward. One second later (b) the fireball has covered the surface of the white dwarf, incinerating carbon and oxygen as it goes. When the fireball completes the transit and collides with itself it triggers an inward propagating shockwave. In (c) the shockwave has compressed the core of the white dwarf and triggered a detonation, which then begins messily unbinding the white dwarf generating the supernova. The entire process lasts less than 3 s and is very asymmetrical

the fireball erupts initially outward and then, in a manner akin to a pyroclastic flow, spreads rapidly outward in all directions, covering the star in a second or so (Fig. 11.2).

As the fireball sweeps radially it collects, heats and incinerates more carbon-rich material from the surface of the star. Within a second or two the fireball's edges collide on the opposite side of the star and violently compress the swept-up fuel. This triggers the detonation required to produce the correct elemental abundances. Most of the interior of the dwarf is then violently shocked by the detonation and explodes, scattering an abundance of iron-rich debris into interstellar space.

This model explains a number of observations that indicate asymmetry not only in the shape of the explosion but in the distribution of elements. For example, Spitzer observations suggest that some Type Ia supernovae remnants contain a very polarized distribution of intermediate mass elements – those that would be produced by the initial deflagration. However, the iron-group elements synthesized in the detonation phase are more uniformly distributed. The GCD model naturally explains this distribution; its rivals do not.

However successful, it shouldn't be forgotten that this is a model. Get a single parameter wrong and the outcome of even the most complex model may differ significantly from reality. Without getting up close and personal to Type Ia supernovae, it may not be possible to precisely match the conditions that give rise to the explosion. For example, Dr. Wolfgang Hillebrandt (Max Planck Institute for Astrophysics, Garching, Germany) used very similar initial conditions but employed a different treatment of the mixing of material within the burning front. The attempts of his group failed to duplicate the detonation. Subtle differences in the input physics produced a very different outcome. Therefore, successful though this model is, it is just that – a model. A combination of further observation, improved computing power and just a hint of luck may finally reveal the inner workings of these surprisingly complex explosions. Conflicting though Hillebrandt's observations initially appear, the divergence in the outcome of the modeling in different runs might explain the diversity in the observed Type Ia supernovae. Rather than a problem, Hillebrandt and other's work may open up avenues allowing astronomers to explore the full scope of these supernovae (Fig. 11.3).

SN 2011fe: A Defining Type Ia Supernova

Supernova SN 2011fe was in many ways a magical event. It was close enough to be observed by backyard observers, and moreover, observations of the explosion were made at very early stages, constraining many theoretical models of these supernovae.

Type Ia explosions were believed to follow the destruction of a carbon-oxygen white dwarf. Naturally, one would have assumed that carbon and oxygen would be present in the debris of these events. However, with few exceptions, these two elements were

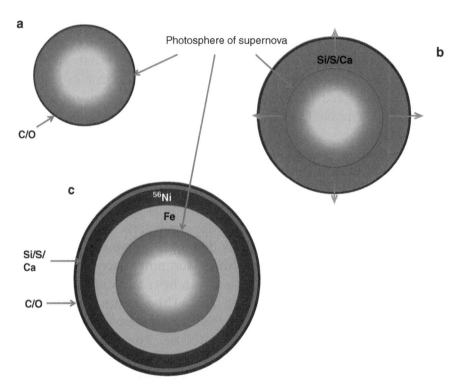

FIG. 11.3 The progressive unveiling of a supernova. (a) Photosphere of supernova illuminates outermost carbon-oxygen layer, giving rise to absorption lines in spectrum. Inner layers obscured by the high opacity of dense, hot photosphere (*deep, graded blue*). (b) Cooling and expansion of outermost layers with shrinkage of photosphere inward. Light from photosphere illuminates inner silicon and sulfur layer. Expansion of supernova debris also dilutes outer layers, allowing light from inner photosphere to emerge. (c) Further shrinkage of photosphere and light from this hot surface illuminates innermost iron-rich layer. If the debris is mixed well, this pattern is disrupted, as was seen in SN 2011fe

difficult to pin down in the expanding debris. Part of the problem was catching a supernova early enough so that these elements were detectable.

In a Type Ia supernova the star produces a wealth of intermediate mass elements deep with the core of the star, including the iron group elements. Nearer the surface, where temperatures remain low, some unburned carbon (and oxygen) fuel is likely. Seeing this carbon skin requires catching the explosion at times early enough so that the photosphere (the radiating surface of the supernova – Chap. 2) is at a suitable depth to illuminate these elements and produce the necessary absorption features. Deeper layers are illuminated as the photosphere retreats inside the expanding cloud of debris.

Early observations of SN 2011fe by Peter Nugent (Lawrence Berkeley National Laboratory) and co-workers identified unburned oxygen moving at 20,000 km/s away from the center of the explosion. This was far higher than was anticipated in most models. At these early times only a fraction of a solar mass of material was present above the photosphere of the supernova. Therefore, it was easy to spot the fast-moving oxygen-rich debris, along with other intermediate mass elements such as silicon, calcium and magnesium. The physical nature of this unburned material was unclear. It could have been in some form of dilute outer shell, or perhaps in the form of nuggets of material mixed into the deeper silicon- and calcium-rich debris. At these early times more strongly ionized silicon and iron was absent, confined to deeper material. However, shortly thereafter iron became visible. This indicated that a substantial amount of mixing was present in the supernova, with elements from deep down being drawn into the upper layers at very early times.

In the context of the available models, the confined gravitational explosion (GCD) seems to produce the desired mix of elements and their velocities. With the initial detonation occurring near the surface of the white dwarf astrophysicists can explain the high velocity of the oxygen in SN 2011fe. Moreover, the rising plume of burning material will bring iron-rich debris towards the stellar surface from the interior, where they can be observed early on.

SN 2011fe was also notable for one other reason. Early observations of the outgoing shock wave by Nugent, and analysis of archival images by Weidong Li (KAIT) ruled out a red giant companion, leaving a main sequence star, a helium star or perhaps a second white dwarf as the likely accomplice in the detonation. Whatever its nature, there was little evidence of its contribution to the light of the supernova, indicating that there was little heating of the star by the outgoing supernova shockwave. Once again, the single-degenerate progenitor model was found lacking.

Conclusions

Automated supernovae searches are revealing a vast treasure trove of events that are helping constrain the nature of these explosions. Type Ia events are no exception. Serendipity has also played a key role in developing and constraining our understanding of these supernovae. For example, the data gathered from SN 2011fe has revolutionized much of our understanding of these events. Data from this explosion eliminated the SD route, exposed high velocity material from the skin of the white dwarf and exposed some of the internal working of the supernova for the first time. Although Type Ia events have been broadly understood for many years new observations are constraining and developing the science in equal measure. Consequently, the field of research into Type Ia events has remained vibrant at both a theoretical and observational level. Glory days.

Finally, observations of Type Ia events are also opening up further cracks in the classification system of supernovae in general. SN 2002ic is a case in point. Returning to genetics, finding a Type I supernova with hydrogen is a little like finding a plant that walks and talks. The classification scheme was built around a perceived singular schism between hydrogen-rich and hydrogen-deficient supernovae. Taking SN 2002ic at face value we now have a hydrogen-rich-deficient supernova that is something of an oxymoron, if ever there was one. This merely serves to underscore the need to overhaul the classification scheme for supernovae. After all, this horse and cart scheme was developed before the motorcar or airplane. In Chap. 13 we encounter "Type .Ia events," which, in terms of their name, is really rather silly. A new, flexible scheme is needed. Suggestions on a postcard, please – or see the final chapter.

12. Are there Super-Chandrasekhar Supernovae?

Introduction

Can a white dwarf, filled with carbon and oxygen, be more massive than the Chandrasekhar limit of 1.37 solar masses? In theory – and this is a very reliable theory – no. At this limit, or indeed somewhat less than this limit, the conditions in the core should turn carbon into iron and other elements, with disastrous consequences. However, in 2003 the Supernova Legacy Survey discovered an unusual Type Ia supernova called SNLS-03D3bb that appeared to break this rule, and not just in a subtle way. SN 2003fg, as it was subsequently renamed, had a peak absolute magnitude of –20.09, requiring a mass of radioactive nickel-56 almost as huge as the theoretical limit for the entire white dwarf. This appeared to be heretical. How could astronomers square this round peg?

Type Ia Supernovae: A Reprise

Type Ia supernovae result from the incineration of an entire carbon-oxygen white dwarf star. The star will have a mass roughly 1.36–1.38 times that of the Sun. During the explosion carbon and oxygen within the star are violently fused so that most of the interior of the dwarf becomes iron, cobalt and, crucially, the radioactive isotope nickel-56. Most of the remainder is transformed into lighter elements such as silicon, sulfur and magnesium. Indeed the defining features of these explosions are the prominent emission line of silicon at 635.5 nm and an abundance of radioactive nickel-56 – roughly 0.6 solar masses. These values are fairly constant for the majority of Type Ia supernovae, although there is a growing army of exceptions, mostly fainter than expected.

The physics underlying the explosions is reasonably well known, and there are considerably fewer variables to consider than in core-collapse events. Simply put, given that these supernovae appear to arise through the same mechanism, the same general rules should apply to each explosion. So, if the explosion originates inside a 1.3682 solar mass carbon-oxygen white dwarf star, the physics of the interior should be fairly uniform from explosion to explosion. Consequently, any Type Ia supernova that deviates from the standard scenario would raise questions about the circumstances leading up to, and the generation of, these explosions.

SN 1991T

SN 1991T was a peculiar supernova in a number of respects. It was brighter than normal with a broader than normal peak. It displayed highly ionized iron early in its spectrum; and the early spectrum was oddly featureless. It was only after day 14 that other spectral features characteristic of Type Ia supernovae gradually appeared.

Carbon was observed early in the spectrum, an element expected on theoretical grounds, but one that should be moving with velocities on the order of 20,000 or more kilometers per second. In SN 1991T this element was moving considerably slower, at speeds more typical of the deeper, intermediate mass elements, such as calcium. The early spectrum was thus an enigma – lacking some of the expected elements, containing others – while the ejecta were moving slower than normal (Table 12.1).

Adam Fisher (University of Oklahoma) and others suggested that this supernova's extra light, as well as some of the peculiarities in the spectrum of the explosion, might be explained by an eruption originating in a white dwarf with a mass well in excess of the Chandrasekhar limit. At the time of publication in 1998 it was unclear whether this was reasonable, or whether this type of peculiar Type Ia event was a rarity with correspondingly peculiar or unique circumstances. At the time relatively few Type Ia events had been recorded, and there was little hard data to compare it to. It would take a further very bright Type Ia supernova to shove the *super*-Chandrasekhar model forward.

Are there Super-Chandrasekhar Supernovae? 277

Table 12.1 The properties of SN 1991T. The key observations are the higher overall luminosity and the slower than expected expansion velocity of the debris. Also significant is the detection of carbon and iron at early times

Ejecta velocity	Ejecta mass/inferred mass of Nickel-56	Element identified from spectrum	Peak absolute magnitude	Spectroscopic features
13,000 km/s measured for Ca layer and 9–10,000 km/s for deeper Si layer; 16,000 for C but lower than expected 21–26,000 km/s	1.14 solar masses	Si, Ca, S, Fe: standard Type Ia supernova; high velocity C early	−20.28	Broad maximum light. Fairly bland, featureless spectrum early on. Notably carbon at earliest times, but with strongly ionised Fe. A more typical Type Ia spectrum from day 14 with Si II line and Fe lines; limited Ca, S and Si in outer layers of blast; Fe dominated later on as seen in other normal Type Ia SN

Most papers assume a larger than normal mass of nickel-56 than standard Type Ia SN, but there is considerable disagreement, including the mass of the white dwarf.

Although the super-Chandrasekhar model seemed possible it arrived against a backdrop of many other competing ideas. Having a white dwarf above the famous mass limit was by far the most contentious and hence least appealing model. Previous studies suggested a very different origin for SN 1991T – the double detonation of a sub-Chandrasekhar limit white dwarf. In this scenario an accumulated helium layer ignited and underwent a detonation. The shockwave from this explosion compressed the underlying dense carbon-oxygen white dwarf core until it, too, exploded.

However, this model failed because it could not generate sufficient nickel-56 to power the light curve. Therefore, the possibility that some Type Ia events might be powered by larger-than-Chandrasekhar masses seemed, perhaps, farfetched. After all the consensus held that Type Ia events occurred when the mass of a white dwarf reached, not exceeded, the Chandrasekhar limit. Could a white dwarf exceed this mass limit, or was there a more prosaic explanation?

SN 2003fg: Too Bright for Its Own Good?

In 2003 SN 2003fg was observed by the Supernova Legacy Survey that was comparable in luminosity to SN 1991T. Like its predecessor the luminosity implied that this eruption must be powered by 1.3 solar masses of radioactive nickel-56. The physics of the nuclear reactions that produce nickel-56 in Chandrasekhar-mass Type Ia explosions struggle to make 0.9 solar masses of nickel-56, regardless of its composition. So, how could a Type Ia supernova produce so much radioactive nickel from a standard white dwarf?

SN 2003fg occurred at a red-shift of 0.244 – reasonably far back into the universe's past, and the stellar population in which it arose appeared to be relatively young. Was there something different about the neighborhood of the explosion that might help explain its peculiarity?

Spectra of the explosion also revealed that the material was moving at *only* 8,000 km/s. This is considerably slower than the canonical value of 11,000 km/s seen in other Type Ia supernovae at comparable times. The light from the explosion also indicated that some carbon had remained unburned, giving rise to a faint emission line in the spectrum.

Although not a first, the presence of carbon in the spectrum of Type Ia events is rare and often rather suspect. Could scientists explain both the sluggish nature of the explosion and its remarkable fecundity in terms of radioactive nickel? Combining the data, Andrew Howell (University of Santa Barbara, California) suggested that SN 2003fg arose in a star with a super-Chandrasekhar mass – resurrecting the original proposition put forward for SN 1991T.

How does a super-Chandrasekhar mass explain the features of these explosions? A Type Ia event is a relatively simple beast. Carbon begins to burn at or near the center of the star. This initially produces a rapidly rising plume of super-hot material. In the most successful models, this erupts onto the surface of the white dwarf within a second of its genesis and pours across the stellar surface. This roiling mass is initially contained by the white dwarf's intense gravity. Within the plume carbon burns to make iron, nickel and cobalt, while around it further material ignites.

As the plume rapidly sweeps across the surface of the dwarf it eventually consumes it before colliding with more material spreading outward in the opposite direction. The collision of this dispersing conflagration converts its kinetic energy into heat. With an additional blast of energy, the flame turns explosive, generating a violent shockwave that drives into the center of the white dwarf – and out into space. Compressed and heated even further, the core of the white dwarf star explodes, sending iron-rich debris out into space.

If the star has even more mass, might the larger bulk of carbon fuel drive a more powerful explosion? Two opposing forces come into play. The greater the mass the star has might in turn cause a greater restraint upon the motion of the shocked gas. This might then cause the eruption to expand more slowly. However, conversely, the greater mass of fuel should lead to a greater production of radioactive nickel-56, which in turn would lead to a brighter explosion. This, in essence, was Howell's conjecture for SN 2003fg. However, as Stan Woosley pointed out at the time, would not the greater mass of nickel-56 lead to a more energetic blast? Perhaps unsurprisingly, given the potential contradictions, it received a fairly mixed reception.

In principle the model for super-Chandrasekhar supernovae (and indeed all Type Ia events) works by virtue of its apparent simplicity. The kinetic energy of these supernovae reflects the initial input energy provided by the radioactive decay of elements – primarily nickel-56. In its simplest determination, the kinetic energy of the explosion is the difference between the input radioactive energy and the binding energy of the star. The binding energy is the amount of energy needed to eviscerate the star. Therefore, a more massive star will require more energy to shred it and that available to expand the remains correspondingly less.

Thus the high luminosity, coupled to the relatively slow rate of expansion, implies that a lot of energy went into expanding the star – and this in turn implies a high mass of the progenitor. Indeed, we have already seen a persuasive argument for SN 2007bi as an explosion driven by pair instability and powered by several solar masses of nickel-56. Yet, here, the supernova debris was thought to be moving relatively slowly. Perhaps the relationship between radioactive power and kinetic energy is not so straightforward (Chap. 8).

Inherently contradictory, some in the astronomical community pointed out that had there been more radioactive nickel synthesized in the explosion, then the energy available to expand the star would have been greater. This would have meant a proportionately greater amount of expansion, rather than a simple increase in luminosity. This is a difficult circle to square. However, this assumption depends on the efficiency of conversion of nuclear energy into kinetic energy – and this is unlikely to be straightforward. If much of the energy from radioactive decay escapes early, then the material won't receive the additional kick from this extra radioactivity. This is particularly true if a substantial fraction of this nuclear fuel is near the surface of the conflagration as the gravitationally confined detonation model might suggest. Consequently, it won't expand as much, and the debris cloud will be brighter.

Loneliness in a Crowd

Were SN 1991T and SN 2003fg simply two lone peas in a pod? Certainly super-luminous Type Ia supernovae are rare, with only a whiff of these being found per decade. Since the discovery of SN 2003fg nearly a decade ago, only three other similar supernovae have been caught: SN 2006gz, SN 2007if and SN 2009dc.

Are there Super-Chandrasekhar Supernovae?

Table 12.2 An overview of some of the features of the four proposed super-Chandrasekhar Type Ia supernovae found in the last decade. These rare explosions have a number of consistent features, including luminosity and the presence of carbon in the early spectra

Supernova	Peak absolute magnitude	Expansion velocity at peak magnitude (km/s)	Presence of un-burnt carbon
SN 2003fg	−20.09	8,000 ± 500	Weak but present
SN 2006gz	−19.74	11,500	Strong
SN 2007if	−20.4	8,500 ± 400	Strong
SN 2009dc	−20.0 (some uncertainty due to intervening dust)	8,000 ± 500	Strong

Indeed SN 2007if was even brighter than SN 2003fg, with a peak absolute magnitude of −20.40. In 1988 another explosion, SN 1988O, may have been even brighter at −21.3. However, the circumstances of this explosion are rather more uncertain, and this result is questioned (Table 12.2).

These recent bright Type Ia supernovae have a number of unusual factors in common, which may help explain why they are so much brighter than other Type Ia explosions. Firstly, they all exhibit slower expansion velocities than other Type Ia events – although SN 2006gz is more marginal. They each have high peak luminosities indicative of the decay of 1.2–1.6 solar masses of nickel-56. And there is evidence of high degrees of ionization in some of the detected elements, which might suggest high temperatures associated with this extra nickel-56. Importantly they also show the presence of substantial amounts of unburned carbon. Finally, they appear to have detonated in low metallicity environments.

Can this evidence shed light on their peculiarly brilliant light? The simplest scenario involves the white dwarf having more than the standard amount of mass, but how to achieve this? Hypothetically, if the white dwarf is accreting matter at a sufficiently high rate accretion could spin up the white dwarf so that it is centrifugally supported against gravity. However, the white dwarf would effectively have to double its mass, reaching more than twice the mass of the Sun prior to detonating.

Could such a mass of material be held up for sufficient lengths of time by mere rotation? Remember the core has to heat up as well as spin up, and these effects are opposed by various processes involving the loss of neutrinos. These cool and contract the core as more mass is added. Given these effects, it seems unlikely that a white dwarf could naturally accrete enough mass to bypass the Chandrasekhar limit by such a large extent without quickly collapsing.

Even if we invoke a common-envelope phase with the white dwarf embedded within another star, it seems unlikely that the dwarf could simply accrete gas fast enough to raise the mass of its core to such gargantuan proportions. Occam can be heard sharpening his razor.... Thus the single-degenerate scenario seems improbable, if not entirely ruled out.

The double-degenerate scenario seems more promising. Here two white dwarf stars merge. If they are already around 1 solar mass apiece then the merger will naturally produce an object – albeit briefly – with sufficient mass to power a super-Chandrasekhar event. More interestingly, as the two white dwarf stars spiral together the less massive of the pair will suffer the slings and arrows of outrageous fortune and become shredded to produce a small, dense, carbon-rich accretion disc around the other, more massive star. When the interior of the latter star finally incinerates, a lot of this shed carbon-rich material will end up caught by the wave of expanding debris and blasted into space. All this unused carbon fuel will be too cool to take part in any further nuclear reactions. Thus, in principle, the double-degenerate route could produce a star capable of generating an ultra-bright, "super-Chandrasekhar" event and one that will display carbon in its spectrum.

Is the environment of these explosions important? Each of these "super-Chandrasekhar" events seems to have occurred in metal-poor environments. Conceivably, a lack of metals may lead to a population of more massive white dwarfs – if the amount of mass lost from their progenitors during the red giant phase is restrained. Reducing mass loss in the parent star will also allow both stars in a hypothetical binary system to cling more closely together. Mass loss reduces the gravitational bond they have on one another and could allow them to spiral apart over time – particularly during a red giant phase. However, if they retain more mass until near the ends of their lives their orbits will remain tight, affording a greater opportunity for each to interact later on, generating a supernova.

Are there more banal explanations? Perhaps the explosion isn't brighter overall than expected, and the extra luminosity comes from observing an asymmetrical explosion? The most successful model for a Type Ia event – the gravitationally confined explosion – certainly is not spherical, nor is it even vaguely symmetrical. There is some evidence for asymmetry in Type Ia explosions. However, invoking the viewing angle seems an unlikely explanation for the observed differences in the luminosity of these explosions.

What about a Type IIn scenario, involving the interaction of the blast wave with circumstellar material? The low photospheric velocities might be suggestive of some form of circumstellar interaction, and a "velocity plateau" in SN 2007if might suggest that the material was decelerated early on in the explosion's history. However, it is difficult to generate light from this mechanism without there being evidence for a dense CSM in the early spectra of each explosion. In particular such collisions may be expected to produce an early UV excess in the detected emission. This is not seen.

In the simplest scenario the majority of the detected luminosity comes directly from the explosion itself. However, whether the single- or double-degenerate scenarios are involved, there should be at least small amounts of un-accreted material surrounding the doomed white dwarf prior to its obliteration. Material naturally finds itself there during accretion – and it would seem difficult to avoid some sort of interaction like this in either scenario. Thus at present, the double-degenerate model seems the most promising route to super-Chandrasekhar events.

Conclusions

Super-Chandrasekhar Type Ia supernovae were another recent addition to the menagerie of explosions. Each animal has been discovered throughout the last decade in large part through the amplification of automated discovery programs. At present these are the rarest of the luminous supernovae, which may reflect an intrinsic scarcity in numbers or some form of selection bias. It is also possible that there are kin out there which are misidentified as different kinds of events and hence overlooked. If these events truly do reflect the destruction of super-massive white dwarf stars, then developing a full explanation will undoubtedly await a better understanding of both the evolution of systems that give rise to Type Ia events and a clearer picture of the dynamics of the explosion itself.

13. The Good, the Bad and the Ugly Thermonuclear Supernovae

Introduction

The basic tenant of the Type Ia supernova paradigm is that these explosions happen when a carbon-oxygen white dwarf gets to (or more precisely gets close to) a tipping point – the Chandrasekhar mass limit. It then ignites its store of inert carbon and oxygen. The resulting explosion decimates the white dwarf, leaving an expanding cloud of radioactive debris dominated by nickel-56 and other iron group elements. So far, so good. Accepting this premise then begs the next question. Will any sort of white dwarf conflagration do, or do different circumstances trigger different explosive – or perhaps non-explosive – outcomes?

We have already seen that a rare few Type Ia events are overly luminous, and this may relate to the detonation of super-Chandrasekhar mass white dwarf stars. Can other situations arise that produce thermonuclear supernovae that are as distinctive as the super-Chandrasekhar events?

Evidence now suggests that there are a variety of possible outcomes from white dwarf mergers, or from the accretion of matter onto a white dwarf from another star. Some of these events result in supernovae, while some merely rejuvenate the white dwarf. Supernova searches have now revealed a large variety of supernovae that do not readily fit into any pre-existing category. Consequently, there now exists a rag-tag collection of explosions with peculiar light curves and even more peculiar chemistry.

Although often difficult to pin down, the evidence is now pointing to these odd supernovae as variants of helium novae, or incomplete Type Ia events. Here, the bulk of the white dwarf is transformed, but it is not decimated in the eruption. Moreover, it appears as though most Type Ia events may be intrinsically super-Chandrasekhar in nature. This follows from a growing understanding of how the merger of two white dwarf stars leads to Type Ia explosions. In most instances such mergers will naturally produce an object, albeit temporarily, that has a mass in excess of the canonical limit rather than one that is fine-tuned with this mass. In essence, then, the nature of the resulting explosion will come down to the total mass of the two coalescing white dwarfs. It is this variable that may set the explosion, whether overly bright or under-luminous. By contrast the single degenerate route cannot produce this range of stellar masses, at least not so easily.

In order to try and bring some semblance of order to this haphazard bunch of explosions, a variety of weird and wonderful events are explored and some sort of explanation for their aberrant behavior is discussed. It's still very early days for these cutting-edge supernovae, and a lot more needs to be learned about how some white dwarf stars explode or transform. Back in 2000 a question was posed to James Kaler, who was then answering questions in *Astronomy* magazine's "Ask Astro" section. "Are there iron core white dwarfs?" At the time the answer was thought to be no. White dwarfs came in three flavors: helium, carbon-oxygen and oxygen-neon-magnesium. It now appears that there is – or may be – a fourth variety, and its formation is the source of some of the most unusual supernovae discovered in recent years. Much is changing; much is being learned. These are exciting times.

R Corona Borealis: Born Again Stars

R Corona Borealis stars, or RCrB stars for short, are downright odd. These are very luminous, yellow supergiants with only the barest whiff of hydrogen (about 5 % by mass); abundant helium; a rich tapestry of carbon compounds, ranging from soot to polyaromatic hydrocarbons; and elements such as fluorine, which are believed to be synthesized in rare environments. They also show abundant elements synthesized by the s-process, a series of reactions whereby neutrons are slowly added onto "seed nuclei" such as iron. These reactions cause a progressive build up of heavier elements, up to lead (Pb-207) in the Periodic Table. These sorts of elements are dominated by isotopes that are predominantly stable but neutron-rich.

This type of variable star shows prolonged periods of high luminosity, interrupted by broad dips caused by the formation of thick, opaque clouds of carbon soot. The lack of hydrogen, and the apparently low mass of the star (less than a couple times that of the Sun), indicate an odd origin. Most of these variable stars are spectral class F or G, but a few are hotter, with spectral classes A to late-B. These correspond to surface temperatures of 5,000–7,500 K for the cooler stars and 8,000–15,000 K for the rarer, hotter stars.

In principle two stellar evolution pathways could produce these objects. In one, an aging AGB star begins its journey to become a white dwarf, with hydrogen fusing in a layer atop the inert helium shell that surrounds the carbon-oxygen core. In a few instances – in perhaps 10 % of these stars – the helium layer becomes deep enough during this phase to begin fusion reactions. Convection then drags the remaining hydrogen deep into the helium burning zone, where it is burned alongside the helium. This produces the s-process elements. Meanwhile carbon is dredged upward into the expanding hydrogen and helium envelope.

These "born-again" stars are typified by Sakurai's Object. This star evolved backwards from a non-descript white dwarf into a luminous red giant, all inside the space of 5 years. The rebirth of Sakurai's Object confirmed some born-again scenarios which suggested that the evolutionary processes should be quite swift. These stars have limited fuel reserves, which would ensure a correspondingly swift transition from one state to another. Thus these stars are expected to return to the white dwarf cooling tract within a couple of centuries of re-birth. However, in contrast to Sakurai's Object, R Corona Borealis stars seem to have more staying power, with lifetimes measured in hundreds to thousands of years. Thus a second formation route for these stars seems likely. What is it?

In 1996 a second model was proposed by Icko Iben (University of Illinois), Alexander Tutukov and Lev Yungleson (both at Institute of Astronomy, RAS). In this merger model R Corona Borealis (RCrB) stars were produced when a low mass helium white dwarf fused with a more massive carbon-oxygen white dwarf. Initial modeling proposed that a white dwarf, composed primarily of helium and in the range of 0.25–0.4 solar masses would be tidally disrupted by its more massive 0.7–0.9 solar mass carbon-oxygen companion. The bulk of the helium would then settle onto the carbon-oxygen white dwarf, and the residual hydrogen, present in both stars, would begin to burn. Helium would also rapidly ignite once its density and temperature became high enough. The large increase in energy would then inflate the helium and hydrogen layer, generating the yellow supergiant star – the RCrB object. Hydrogen burning would be aggressive, and, once again, vigorous convection would ensure that the products of fusion were transported outward through the bulk of the envelope.

Convection is crucial in generating the unusual chemical composition of these stars. The s-process elements are believed to be produced by an off-shoot of the CN cycles (Chap. 1). Neutrons are released from one reaction involving carbon-13, while the carbon, in turn, originates through helium fusion. The structural configuration within RCrB stars naturally explains the abundance of these elements in these stars. Similarly fluorine, a very fragile element, can be created by RCrB stars. In most scenarios, fluorine is produced by hot bottom-burning (HBB). HBB requires that the products of hydrogen fusion are quickly removed from the site of their creation. At very high temperatures the CNO cycles generate fluorine, but the fluorine is also readily destroyed at these temperatures. In order to survive, the freshly minted fluorine must be dragged away to cooler climes, where it is stable and free from the risk of further reactions that can destroy it. The structure of the RCrB star, with its low mass and relatively cool envelope, but a very hot core, around which hydrogen and helium are burning, would be a natural site for the creation of this reactive element.

Over the course of a few thousand years, RCrB stars should exhaust their supply of helium fuel and begin the process of returning to the white dwarf cooling tract, somewhat more massive, substantially hotter and, at least briefly, reinvigorated. The population of hotter and rarer RCrB stars may simply be a manifestation of the final stages in the evolution of these stars. As the supply of helium wanes and the stars begin to contract once more, they heat up. This finally transforms the RCrB star into a smaller, and even hotter, so-called extreme helium star.

Extreme helium stars are hot B- or O-class objects that are burning their residual helium in a shell above their hot, dense carbon-oxygen core. These stars are smaller than the Sun, but with a similar luminosity, comprising a hook between the end of the horizontal branch and the white dwarf tract on the HR diagram. It is believed that many of these objects represent the contracting, fading RCrB star as it slowly transforms one final time into a hot white dwarf.

Mergers of More Massive White Dwarf Stars

What happens if we start the merger with two more massive white dwarfs, both with a carbon-oxygen composition and roughly 0.9 solar masses each? Rudiger Pakmor and colleagues (Institute for Theoretical Studies, Heidelberg) presented modeling of such binary systems in the journal *Nature* in 2011, and subsequently, a broader range of masses in a follow up paper. The key finding was that such mergers produced sub-luminous Type Ia supernovae; explosions dominated by slower-moving ejecta. The models made a fairly convincing match for a population of events that included SN 1991bg. Fainter than average SN 1991bg-like events are fairly common, making up approximately 10 % of observed Type Ia explosions. Indeed it was their relative abundance, coupled to their unusual properties, that made them the ideal target for computer modeling.

Pakmor and colleagues showed that when two similar mass white dwarf stars began to merge, the more massive of the two stars began to distort, then fatally disrupt, the less massive but larger white dwarf. Any helium present is rapidly consumed in nuclear reactions, while the shredded carbon-oxygen core drops violently onto the surface of the denser, more massive star. With a combined mass of more than the Chandrasekhar limit, the resulting body rapidly destabilizes, and the carbon-oxygen core ignites (Table 13.1).

Modeling suggested that in the ensuing maelstrom approximately 0.1 solar masses of radioactive nickel-56 was produced: considerably less than that synthesized in the super-Chandrasekhar explosions considered in the previous chapter. It is also a sixth the amount typically synthesized in standard Type Ia events. So, how does this model, and those that were produced subsequently,

Table 13.1 The properties of the sub-luminous Type Ia supernova, SN 1991bg

Supernova	Overview of properties	Ejecta velocity	Ejecta mass/mass of Ni-56	Elements identified from spectrum	Peak absolute magnitude	Spectrum	Best model
SN 1991bg	Red optical colour at peak but normal (blue) late on; light curve rise and fall 40 % faster than normal; brief photospheric phase – transition to nebular phase 2 weeks post-max	Lower than typical Type Ia (but not a dramatic difference)	0.1 solar masses Ni-56/1.3 solar masses ejected); an earlier paper by Robert Kirshner suggested 0.5–0.8 solar masses of ejecta	Ca, S, Si, Fe at late times, Ti, Mg and other intermediate mass elements; possible C and O early (Ca and Si strong near maximum)	−17	Depressed spectral region between 410–440 nm with few absorptions; lower luminosity consequence of low ejecta mass – less trapping of gamma rays from co-56	Merger of two ~0.9 M CO white dwarfs (Pakmor, Nature 2011)

recreate the spectrum of SN 1991bg? With a lower than average amount of radioactive nickel-56, the amount of energy available to drive the explosion is reduced. Moreover, the mass of debris the detonation has to shift is greater; therefore, the available energy is shared between a proportionately greater mass in the model. This means that the velocity of all these debris velocity is slashed, neatly explaining observations.

However, this raises the question of why a larger than Chandrasekhar mass produce so little nickel-56? Observations of SN 2005bl – another prototype of this class of under-luminous explosion – indicated that silicon and iron were thoroughly mixed in the explosion – all the way through to its heart. Although there is some evidence of mixing, in normal Type Ia events there is a progressive increase in the masses of the elements synthesized as you move inwards through the explosion debris. This is a natural consequence of the inside-out nature of the explosion that drives these supernovae. However, in the sub-luminous explosions these elements were thoroughly mixed, suggesting a much lower and more homogeneous range of density throughout the star. The higher than normal degree of homogeneity in the debris, in turn, suggested that a violent merger process had shaken up the star in its entirety (Fig. 13.1).

Pakmor's modeling of the fireball showed that the explosion quickly became supersonic (a detonation) at early stages. This prevented the formation of a temperature gradient within the 3-billion-degree fireball. This in turn ensured that burning was both incomplete yet fairly homogeneous. Overall the energy released was similar to a standard Type Ia supernova. However, it is worth reiterating that the available energy was shared between a larger mass of material, neatly explaining an explosion that was dimmer, redder and slower in its expansion than average. Let's say you follow and accept this explanation. Why, then, were SN 1991T and SN 2003fg more than a little ostentatious, while SN 1991bg considerably less so? After all, we are suggesting that they are all representatives of the super-Chandrasekhar fraternity.

SN 1991bg was observed in an elliptical galaxy, while SN 2003fg was clearly associated with a young stellar population. Moreover, the mechanism proposed for SN 1991bg involved the merger of two roughly equal mass stars. Perhaps the progenitors

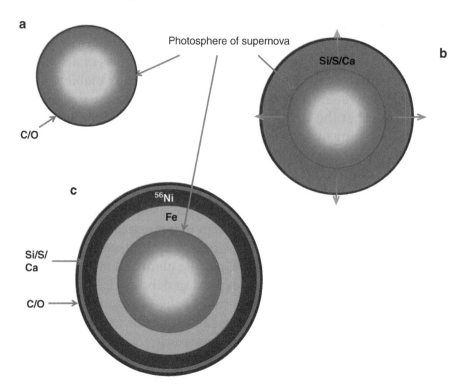

FIG. 13.1 The progressive unveiling of a supernova. (a) Photosphere of supernova illuminates outermost carbon-oxygen layer, giving rise to absorption lines in spectrum. Inner layers obscured by the high opacity of dense, hot photosphere (*deep, graded blue*). (b) Cooling and expansion of outermost layers with shrinkage of photosphere inward. Light from photosphere illuminates inner silicon and sulfur layer. Expansion of supernova debris also dilutes outer layers, allowing light from inner photosphere to emerge. (c) Further shrinkage of photosphere and light from this hot surface illuminates innermost iron-rich layer. If the debris is mixed well, this pattern is disrupted, as was seen in SN 2011fe

of SN 2003fg were not. This suggests a number of possible scenarios that could account for the two very different outcomes. In the first scenario the progenitor of the super-Chandrasekhar and overly luminous SN 2003fg was associated with the merger of two younger and, conceivably, overall more massive white dwarf stars. We already know that the star that blew up consisted of 2.1 solar masses (Chap. 12) to produce 1.3 solar masses of nickel-56.

Scenario two involves two very unequal mass white dwarfs, with the more massive of the two derived from a particularly massive progenitor star. In a third scenario, the progenitors of SN 1991bg are older and have cooled down more significantly. In a fourth scenario, SN 1991bg may have resulted from a more protracted merger of two stars, with more substantial cooling of the debris prior to detonation. Finally, SN 2003fg could have been the outcome of a single-degenerate event, while SN 1991bg was derived from a double-degenerate scenario.

If we consider the statistics for these events we run into a few problems. For starters the observed population effect could be misleading. Identifying a galaxy with a young stellar population does not necessarily mean that SN 2003fg originated in this young, and hence more massive, stellar system; it merely means that it was found in the vicinity of young stars. It could well be an older binary system that is passing through the spiral arm of its host. Population studies in geographical or biological situations have repeatedly been found wanting when proper controls are absent. It is easy to misjudge the age of a stellar system when it is viewed from afar, as resolution is usually poor and there is no direct evidence of the age of the progenitor. Caution must be the watchword.

Further modeling is underway to shore up the conclusions of these scenarios and to improve the fit to observed sub-luminous Type I a supernovae. However, at present the introduction of exquisite computer modeling is finally revealing some of the secrets of some of the more peculiar and challenging Type Ia events.

AM CVn Systems

In an era of austerity, cutbacks mean that not everyone achieves their full potential. We hoard, fearful of the future, cutting back on all but the bare essentials. Some white dwarf stars find themselves in a similar pickle, with meager fuel supplies and erratic outputs. No more so than those involved in novae.

In a standard nova explosion a hydrogen-rich star, often a low mass main sequence dwarf, donates a drizzle of hydrogen onto the surface of the carbon-oxygen (or occasionally oxygen-neon-magnesium) white dwarf. Once the mass reaches a millionth or so that of the Sun, the hydrogen ignites. Incomplete burning of hydrogen by the CNO cycles causes a buildup of fusion products that are rich in protons. These proton-rich isotopes then decay through the release of positrons that then annihilate with their electron counterparts. Widespread annihilation of these particles causes a sharp increase in the temperature of the burning layer. This, in turn, accelerates the rate of nuclear fusion until the reactions become explosive. The entire hydrogen-burning layer is blasted off into space, causing the star to brighten 100,000 times its original value. This is the nova.

It is expected that nova explosions remove almost all the hydrogen-burning layer, along with an eroded portion of the underlying white dwarf. As such, the mass of the white dwarf should decrease over time.

In a few rarer cases helium is the fuel, not hydrogen. The process of the outburst is somewhat different, but the outcome is similar – a rapid rise in brightness once the fuel is ignited. These AM CVn variable stars, named after their prototype, AM Canum Venaticorum, undergoes periodic brightening. Each time helium ignition triggers a massive inflation of the helium-burning shell around the core of the low mass carbon-oxygen white dwarf star.

In principle there could be three types of companion to the carbon-oxygen white dwarf in these AM CVn systems: a helium burning helium star; a helium white dwarf; or a post-main sequence star that has lost its hydrogen envelope through some form of disastrous interaction with its companion (Fig. 13.2).

AM CVn binary stars have very short periods by virtue of their small, dense stellar components. These must be close to one another if mass is to be accreted by the white dwarf from the companion. In all observed AM CVn binaries, the two stars orbit their mutual center of gravity in less than 70 min. The most evolved of these systems have helium white dwarf companions with very low masses – perhaps lower than 0.1 solar masses. These pathetic remnant stars have reached this withered state through the continuous accretion of their mass by their more massive white dwarf companion. At these very low masses and relatively wide separations the rate of transfer of mass to the white dwarf is minimal. Contrary to what you might expect, a very low rate of accretion may have profound consequences for the fate of the white dwarf.

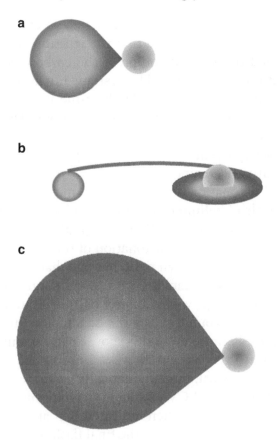

FIG. 13.2 Three possible AM CVn binary systems. (a) In this system a helium-burning helium star is robbed of some of its mass by its carbon-oxygen white dwarf companion. (b) Here, a low mass helium white dwarf loses mass through a stream of matter via an accretion disc onto the more massive white dwarf. (c) Finally, an evolved post-main sequence star, which has lost its outer hydrogen envelope, is the helium donor star. In most observed cases AM CVn systems are of the type shown in example (b) and these work best as an explanation of Type .Ia supernova

Type .Ia Supernovae: Descendents of AM CVn Binaries?

In the mid-1990s work by James Liebert, David Arnett and Willy Benz (all of the University of Arizona) suggested that, in some instances, explosive ignition of helium might be responsible for some Type Ia supernovae. Published in the conference proceeding of the "Icko-fest" *Advances in Stellar Evolution*, Liebert and colleagues tentatively suggested that the ignition of helium under degenerate conditions could compress the carbon-oxygen core to the point that it too detonated. The systems involved were the AM CVn binaries mentioned above. At the time it was thought that these Type Ia events might be under-luminous. However, aside from that, it really wasn't clear just how these explosions might appear until several years later (Table 13.2).

As computing power increased and more work was done, a model for these shell detonation emerged. In the most evolved AM CVn systems, a drizzle of helium can continue to feed the carbon-oxygen white dwarf. Lars Bildsten and Ken Shen (both at the University of California) suggested that the detonation of the helium layer would produce a supernova in its own right, but one that was dimmer than normal and also evolved rapidly. The power source would be the chromium-48, an isotope with a shorter half-life than nickel-56, but also one produced in abundance by the fusion of helium under very dense conditions. Because these supernovae packed roughly 1/10 the power of a typical Type Ia event, they were christened .1a explosions – a somewhat whimsical, if mathematically accurate, name.

So how do these AM CVn systems generate supernovae? Modeling work by Bildsten and Shen indicated that after perhaps ten or so regular novae-like shell flashes the final flash might have a rather different outcome. The interval between outbursts would steadily increase as the amount of available fuel decreased. Consequently, by the time the final layer of helium had accumulated it would have become thoroughly degenerate and relatively massive, its structure matching that of the underlying carbon-oxygen core. However, when the critical mass of helium was finally accreted, ignition of this degenerate layer would lead to a very explosive outcome. Helium would burn through carbon and oxygen to make intermediate mass elements such as titanium, calcium and chromium-48. The latter would be particularly important in this explosion.

Chromium-48 is unstable and decays through the release of positrons with a half-life of a little over 21.5 h. The positrons can provide further heating, accelerating the rate of any nuclear reactions, but primarily providing a lot of energy to heat and illuminate the expanding cloud of debris ejected from the surface of the white dwarf. Chromium-48 also makes a unique power source for these

Table 13.2 Some of the properties of the peculiar supernova SN 2002bj

Supernova	Overview of properties	Ejecta velocity	Ejecta mass/mass of Ni-56	Elements identified from spectrum	Peak absolute magnitude	Spectroscopic features	Best model
SN 2002bj	Fast evolving (5–10 days to peak – 3 times faster than typical Type Ia); decayed 1 mag in 5 days; 4.6 mag decline in 18 days	8,400 km/s but fall quickly to 2,000 km/s 3 weeks later. Possibly much higher earlier as peak missed	0.15 solar masses with uncertain contribution from Ni-56	Si, S and, notably, unburnt C and He present alongside Fe; intermediate mass elements like Ca	−17	Si, S present (as in normal Type Ia supernova) implies C and O burning	Low mass He-shell detonation on low mass carbon-oxygen white dwarf (Lars and Bildsten 2007)

Type .Ia events. In core collapse, pair-instability and thermonuclear supernovae, the principle power source is the decay of radioactive nickel-56 through cobalt-56 to iron-56; each step is accompanied by the release of a positron. With a half-life measured in days to weeks, this short decay chain illuminates a standard supernova for months to years. However, if there is nickel-56 in the debris of a Type .Ia event it is limited to less than 0.1 solar mass. The principle radionuclide must be chromium-48, with its much shorter half-life. With this restricted supply of energy these shell-detonation supernovae have correspondingly short lifetimes.

What would this sort of explosion look like? Well, short models produced an unusually brief and somewhat dim thermonuclear supernova. Table 13.2 describes some of their distinctive key features. Most notable is the peak absolute magnitude of –17 to –19, making them nearly an order of magnitude dimmer than a typical Type Ia supernova. Also of note is the time the supernova takes to brighten to peak luminosity and then fade. In a typical Type Ia event, the supernova brightens over 20 days or so, while a Type .Ia event will do the same in 5–10 days.

Astronomers hit pay-dirt in 2002 when an unusually weak but rapidly evolving Type Ia event was spotted in a distant galaxy. SN 2002bj fulfilled many of the expectations of a supernova with some form of shell detonation powered by chromium-48. Dovi Poznanski (University of Tel Aviv), Alexi Filippenko (University of California, Berkley) and others analyzed the spectrum of this peculiar supernova. The explosion debris was free of hydrogen, indicating that it was a Type I event. However, the brief and diminutive nature of the supernova meant that it didn't fit the bill of any known Type I supernova.

Was it the first discovery of the coveted Type .Ia explosion? In line with the results of the modeling work the peak absolute magnitude of the supernova was considerably dimmer than normal Type Ia events, and the light curve rose and declined within around 20 days of the event. In line with expectations for a .Ia event, the rapid rise and fall implied a low mass of ejecta and a very limited mass of radioactive nickel-56. Indeed the power source was likely chromium-48 or titanium-44, given the shape of the light curve. Later work by Hagai Perets identified further supernovae that appeared to share a kinship with SN 2002bj.

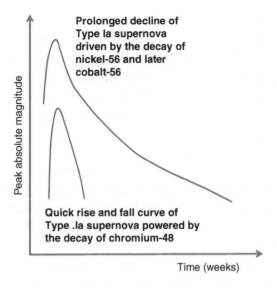

Fig. 13.3 Cartoon illustration of the differences between Type Ia supernovae and their abbreviated relatives, the Type .Ia events

Type .Ia supernovae might not be the last call on the erupting AM CVn system. A final, unique feature is that they might just manage a second supernova. If the orbit of both stars isn't destabilized by the first explosion, the companion might just deliver enough fuel for another go. However, this depends on the impact of the supernova on the carbon-oxygen white dwarf. In a nova, the much less energetic explosion strips material from the dwarf. Consequently, the more powerful supernova might just inflict sufficient damage on both stars to prevent a recurrence, or propel the carbon-oxygen dwarf off into interstellar space away from its stripped partner. It is even possible that the core of the white dwarf is disrupted in these explosions. A combination of modeling and further observation should reveal the intimate details of these diminutive supernovae (Fig. 13.3).

Record Breaking Type .Ia Supernova or Something Else?

SN 2010X was discovered by David Rich of Hampden, Maine on February 7, 2010, and subsequently detected and reported by Mansi M. Kasliwal using the Palomar Transient Factory facilities a few days later.

Superficially, SN 2010X shared a lot in common with SN 2002bj – the prototypical Type .Ia event (Table 13.1). The peak absolute magnitude, the mass of ejected material, and the limited amount of nickel-56 all pointed to the same mechanism – a thermonuclear explosion on the surface of a low mass white dwarf star. So far, so good. However, there were a couple of notable differences that may point to a different origin for this and other, related, supernovae (Table 13.3).

The material in SN 2010X expanded somewhat faster than in other Type .Ia candidates. Although less energetic than a typical Type Ia supernova, the enhanced velocity suggested a somewhat higher energy reserve. The spectrum also revealed one of two possible elements unique to this explosion: helium or aluminum. The distinction between these two is important. Although helium was absent in the spectrum of SN 2002bj, its presence might not be unexpected for an explosion powered by the detonation of a helium shell. However, aluminum is ruled out in Type .Ia models and should be confined to core-collapse supernovae.

SN 2010X was located in a star-forming region that would permit a massive star origin for this event. However, what sort of core-collapse supernova might give rise to such a puny explosion? The collapse of a particularly massive star into a black hole, accompanied by a wimpy venting of the dregs of the star might be one possibility. However, more likely alternative scenarios involve the formation of a neutron star from an intermediate mass progenitor through electron-capture. Or perhaps the accretion-induced collapse of an oxygen-neon-magnesium white dwarf star. The latter two routes would neatly explain the presence of aluminum – if confirmed – and perhaps the low luminosity of the resulting supernova. It is unclear whether such routes could produce the observed rapid rise and fall of the light curve, and this would clearly require further modeling. Intuitively, one feels that the Type .Ia route works best, and the possible detection of the troubling aluminum in the spectrum may be spurious and will go away with further investigation.

Of note it is suggested that two earlier supernovae, SN 1939C and S Andromeda, were also Type .Ia events. However, the late light curves of these events are consistent with the decay of cobalt-56, rather than an element with a short half-life. This raises the possibility that in some Type .Ia events the core of the carbon-oxygen

Table 13.3 Supernova SN 2010X and its unique properties

SN	Light curve	Ejecta velocity	Ejecta mass	Principle elements	Maximum light	Spectroscopic features	Possible progenitor
SN 2010X aka PTF 10bhp (The Type .Ia record holder for speed of evolution)	Very fast evolution (the light curve "half-life" was five days Lower peak than other .Ia events with peak total energy of 1.7×10^{43} J	12,000 km/s falling to 9,000 km/s	0.02 solar masses nickel-56; with an overall mass of 0.16 solar masses	Carbon, Oxygen, Magnesium, Silicon, Titanium, Iron (plus Helium or Aluminum)	−17	Similar to Type .Ia supernova SN 2002bj but with the peculiar elements helium or aluminum	Massive CO white dwarf (1.2 solar masses) with 0.05 solar mass helium-shell. If Al present, accretion-induced collapse (A.I.C.) of ONeMg white dwarf if aluminum present. If Helium is present then He-shell detonation – Type .Ia event

white dwarf might be disrupted, rather than the explosion limiting its impact to the outer helium-rich layer. In such a scenario, the helium layer detonates and lifts off into space, scattering an abundance of intermediate mass elements, such as calcium and titanium. However, in some of these explosions the shockwave is sufficiently powerful to instigate a full detonation of the core. In these more powerful blasts some radioactive nickel-56 is synthesized either in the overlying, shallow helium layer or through a partial disruption of the underlying carbon-oxygen core. If the latter is involved in these explosions, its role must be fairly limited, as little carbon or oxygen appears to be involved in the fusion reactions that occur in the explosion.

Type .Ia explosions are the latest addition to the zoo, and there is plenty of time to flesh out the details of these supernovae, including the extent to which the underlying core is consumed in the blast.

Do Some Type Ia Supernovae Explode Without Detonation?

In a Type Ia supernova, the bulk of the radioactive nickel-56 and the other iron group elements are produced in a detonation, a supersonic conflagration that consumes the core of the star. The most successful models of these supernovae involve some level of deflagration ahead of this destructive final transformation. However, a few rare Type Ia events have ejecta that are deficient in the iron group elements typical of these blasts. What's sets these explosions apart from their siblings?

SN 2005hk was one such supernova. The spectrum was peculiar, lacking a prominent silicon emission line characteristic of Type Ia events, and even more unusual was the presence of iron at all levels in the explosion. If you recall iron is normally confined to the slowest-moving layer that is exposed last as the star disembowels itself. The presence of it throughout the explosion implies unusually good mixing or a very homogeneous and uniformly dense explosion (Table 13.4).

The yield of nickel-56 was also limited (0.22 solar masses), and the speed at which the matter moved away from the center of

Table 13.4 The properties of SN 2005hk – the first detected deflagration-only Type Ia events?

SN	Light curve	Ejecta velocity	Ejecta mass	Principle elements	Maximum light	Spectroscopic features	Possible progenitor
SN 2005hk (SN 2002cx-like)	Lower peak and faster decline; sharper initial rise than typical Type Ia	7,000 km/s (low) and 6,000 km/s for SN 2002cx	0.22 M of Ni-56 for 2005hk (low) and possible high mass of ejecta	Fe and later Si evident; very equivocal evidence for O; no evidence for C; low Si may result from clumpy deflagration as seen in models	−17.9 (2002cx and 2005hk)	Blue continuum early; no Si 635.5 line; but Fe III absorption; after day 11 Si II feature slowly appears; later spectra reveal sharp sharp emission and absorption features not usual to Type Ia. Sn 2002cx and 2005hk spectra at day 11 onwards nearly identical	Deflagration only explosion in CO white dwarf

the blast was also low, at 7,000 km/s at its fastest. The ejecta contained evidence of unburned oxygen. Overall the supernova was under-luminous, showing a faster than normal rate of decline following a somewhat faster rise to peak. Finally, some of the spectral features implied that the total mass of ejected material was rather high, perhaps indicating a close relationship with SN 1991bg. So, did SN 2005hk result from the merger of two white dwarf stars with a total mass greater than the Chandrasekhar limit?

Although no one model could really explain all the features of the supernova, the most successful involved a failed detonation. In this model the supernova began as normal, but for unknown reasons the later detonation stage failed, and the star merely turned itself inside out, scattering the smaller mass of radioactive elements into space. A model involving a collision of two carbon-oxygen white dwarfs might do the trick. In this model, drawing on the work of Pakamor – who looked at under-luminous SN 1991bg-like explosions – two white dwarf stars merge, and the fuel in their hearts ignites. The collision produces a lot of low density material that burns incompletely when ignited. Instead the thermonuclear flame merely burns through the fuel, producing much intermediate mass elements but only limited amounts of iron group elements.

The main problem with this idea is the lack of evidence for unburned carbon in the ejecta – and only limited amounts of unburned oxygen. One might expect more if the explosion was a true failure. However, this assumption rests uneasily on a weak understanding of the processes occurring in Type Ia supernovae, and hence this presumption may well be flawed. Additionally, observations of SN 2011fe imply that carbon and oxygen may only be seen at very early stages in the explosion and may simply be obscured at later times.

In Chap. 11 we saw that modeling work by the Chicago FLASH group indicated that some Type Ia supernovae fail. In these failed explosions there are multiple ignition points rather than one when the Chanrasekhar limit is approached. Because multiple flames are set off, the star as a whole expands and then fails to detonate. Much of the carbon fuel is consumed within the star as it is in normal Type Ia events. However, in some of these failed detonations the iron produced remains bound to the white dwarf. Thus little material is ejected, and what there is moves at low velocities. It might just be that explosions such as SN 2005hk are failures, champagne bottles that failed to pop. Returning to the question posed to James Kaler in 2000, perhaps some white dwarf stars do have a heart of iron. These iron-core white dwarf stars are the failures that persist after they fail to explode.

Whatever the explosion mechanism is, the explanation must fit seemingly contradictory data. SN 2005hk is not alone. A few other recent supernovae, such as SN 2002cx, SN 2003gq, SN 2005cc and SN 2005P, also share many of the features seen in SN 2005hk. Mark Phillips of the Las Campas Observatory, Chile, estimated that as many as 5 % of all Type Ia supernovae may bear the same unusual characteristics, and this implies that the mechanism is fairly common. Some form of merger, perhaps involving two white dwarfs with very different masses, may lie at the root of this sub-class.

SN 2008ha: A Cousin of SN 2005hk?

Discovered on November 7, 2008, SN 2008ha was notable for a number of reasons. Alongside lead investigator Tim Puckett (Puckett Observatory World Supernova Search), the supernova was discovered by the then youngest supernova sleuth on record – Caroline Moore, aged 14. Although Caroline soon lost her title of youngest discoverer to 10 year old Kathryn Aurora Gray, this supernova still holds the title for the faintest supernova recorded. With a peak absolute magnitude of −14.2, this is only marginally brighter than the typical supernova imposter event from an LBV (Chapter 3). The debris from SN 2008ha was extremely rich in calcium, and the outward velocity of the explosion was anomalously low at 2,000–2,300 km/s. Whatever the mechanism was that generated this explosion it was weak to say the least. The great eruption of Eta Carinae ejected material at more than twice this speed.

Spectroscopically, this supernova resembled a class of weak thermonuclear-like events classed as SN 2002cx-like. However SN 2002cx, despite a similar spectrum, produced a supernova with debris expanding at more than twice the speed of SN 2008ha. Furthermore, the absolute magnitude of SN 2002cx peaked at −17, far higher than SN 2008ha. The presence of a fairly bland early spectrum in SN 2008ha was indicative of a dense and opaque fireball, followed by the presence of iron, silicon and oxygen. This favored something akin to a weak Type Ia event. Indeed the spectrum indicated iron early on, suggesting that it was present throughout the eruption cloud rather than, as is more normal, confined to deeper layers that are exposed later on. So, what sort of eruption produces very slowly expanding but well mixed ejecta that is rich in calcium and shows unburned oxygen? (Table 13.5)

A number of rival explanations were published. First up Stefano Valenti (Queen's University, Belfast) suggested that this explosion was a weak core-collapse event. The team concluded that the faintness of the explosion, coupled to the slow-moving ejecta, argued for a failed core-collapse supernova. A preferred model by Valenti had the explosion originating in a particularly massive star, weighing in at 40 or more times that of the Sun.

In Valeti's model the failed supernova led to the formation of a black hole, and the resulting blast wave, sapped of most of its energy, then limped outward at the low observed velocities. The supernova took place near the heart of a barred irregular galaxy, much like the Large Magellanic Cloud. The presence of massive stars in such galaxies would permit this explanation.

Valenti then went on to suggest an alternative explanation, that SN 2008ha was an electron-capture explosion (Chap. 7). Some models show that electron-capture either within the core of an SAGB star or within an ONeMg white dwarf star can produce weak explosions. However, most models of electron-capture explosions in SAGB stars show energies far greater than those seen in SN 2008ha. Moreover, there was no evidence for hydrogen in this blast, or any of the other SN 2002cx-like supernovae. If not here, the absence of hydrogen rules out the SAGB route in all of the related supernovae.

Alternatively, either accretion-induced collapse of an ONeMg white dwarf or the merger of one such dwarf and a lower mass helium white dwarf might do the trick. Indeed the latter route might produce the right mix of helium-contaminated ejecta with a broad distribution of intermediate mass elements seen in the explosion.

More recently Ryan Foley (Harvard-Smithsonian Center for Astrophysics) released a paper explored these scenarios, ruling out most of them. Included in this demolition derby were the

Table 13.5 A comparison of SN 2005hk with SN 2008ha. Although there are suggestions that both supernovae erupted following the same evolution, there are many differences. For example, SN 2008ha showed much slower expansion velocities than SN 2005hk

SN	Light curve	Ejecta velocity	Ejecta mass	Principle elements	Maximum light features	Spectroscopic features	Possible progenitor
SN 2008ha (SN 2002cx-like)	Fast evolving and very dim	2,300 km/s (2002cx was 5,000); kinetic energy 10^{48} erg	0.0029 M Ni-56; from 0.1 to 0.5 M total ejecta (0.2–0.7 M in Foley paper)	Calcium and titanium-44	−14.2	Fe, O, Si, Ca (rich); Ti; presence of silicon argues in favour of Type Ia-like mechanism... Fe well distributed in ejecta; (Similar to bright SN 1991T early on); He likely	He-donor or a deflagration only SN; Alternatively, Accretion induced Collapse or the death of very massive star – a twilight SN Pastorello/Valenti/
SN 2005hk (SN 2002cx-like)	Lower peak and faster decline; sharper initial rise than typical Type Ia	7,000 km/s (low) and 6,000 km/s for SN 2002cx	0.22 M of Ni-56 for 2005hk (low) and	Fe and later Si evident; very equivocal evidence for O; no evidence for C; low Si may result from clumpy deflagration as seen in models	−17.9 (2002cx and 2005hk)	Blue continuum early; no Si 635.5 line; but Fe III absorption; after day 11 Si II feature slowly appears, later spectra reveal sharp emission and absorption features not usual to Type Ia	Deflagration only explosion in CO white dwarf

degenerate shell supernovae: Type .Ia events. These were simply too energetic to explain the features of SN 2008ha and its similarly weak siblings. The most likely survivor of this process of elimination was the accretion of helium onto an ONeMg white dwarf or a pure deflagration supernova, as was suggested for SN 2005hk. With only a sub-sonic blast to drive the explosion, most of the debris expanded slowly in line with observations.

In the Foley 2013 paper the white dwarf is paired with a helium-star: an object with a mass and diameter similar to the Sun. This star is fusing helium to carbon and oxygen in its core. The bulk of the helium-star is non-degenerate and, therefore, the separation of the two stars is greater than in AM CVn systems. The quantity of helium donated to the white dwarf is also greater than seen in Type .Ia progenitor systems, at around 0.1 solar masses. In Foley's model the relatively massive helium shell undergoes a deflagration which then ignites and partly consumes the underlying carbon-oxygen core of the white dwarf. The sub-Chandrasekhar mass of the dwarf prevents the flame getting out-of-hand and detonating the entire star.

However, the chemistry of these supernovae is rather mixed and not all the features match deflagration-only models. For example, Bo Wang (Chinese Academy of Sciences), published their model of the SN 2002cx-like events. Wang's group appeared to follow the *detonation* – or supersonic ignition – of a non-degenerate helium layer atop a relatively massive carbon-oxygen white dwarf. In Wang's model the detonation of the helium layer compresses, heats and detonates the underlying carbon-oxygen core of the white dwarf. In essence they pursue a model for standard flavour Type Ia supernovae that was rejected in the 1990s. A difference in the mass of the sub-Chandrasekhar carbon-oxygen stars explains the range in the observed luminosity of each explosion. Higher mass white dwarfs have more carbon-oxygen fuel hence brighter explosions. The low overall mass of the carbon-oxygen core then explains the low nickel-56 yield and the low velocity of the debris.

Adding the hypothesis from Wang's paper we have a total of three potential models for these explosions: the deflagration of a pair of merging carbon-oxygen white dwarfs; the deflagration of a carbon-oxygen white dwarf that has accreted a non-degenerate layer of helium; or finally, the detonation of a massive helium shell atop a sub-Chandrasekhar carbon-oxygen white dwarf. Of the three the second is the favoured on at present. More data may change this and the sheer diversity of these events may suggest more than one model is correct.

Population studies favour the deflagration of a fairly massive carbon-oxygen white dwarf that has accreted helium from a non-degenerate helium star. In their 2013 paper, Foley and colleagues identified 25 SN 2002cx-like events which are confined to young galactic populations. None are identified in early (elliptical) galaxies where stellar populations are predominantly old. The distribution favours more massive progenitor binary systems and inherently more massive white dwarf stars. Given the growth in the number of these events and the apparently unique population of progenitors, astronomers are now classifying them as *Type Iax*. This further branch of the classification tree adds yet another label which poorly fits with the pre-existing scheme. Surely, it's time for an overhaul? Chapter 15 explores this growing problem in more detail.

How do models of stellar conflagration support these ideas? Computer models of Type Ia events produce very mixed outcomes from standard detonations to pure deflagrations, depending on the precise circumstances of the star before explosion. Modelled supernovae range in power, from ones with enough energy to destroy the white dwarf, to much weaker events, that might be insufficiently powerful to unbind it. Might SN 2008ha be the first example of a failed thermonuclear supernova? Modelling by George Jordan, Hagai Perets and colleagues working with the facitlities at Chicago University's FLASH Centre suggested that this explanation may hold true for SN 2008ha and other 2002cx-like events. Thus SN 2008ha may be a relative of SN 2005hk – only far weaker in nature. Not so much a champagne bottle that failed to pop, rather a warm bottle of soda left standing with its lid off, which someone then shook.

More powerful but related explosions, such as SN 2002cx, might simply involve more of the whole star in the fireball, rather than just the overlying helium layer, or more massive white dwarf stars. Indeed Wang suggests variation in the mass of the white dwarf progenitor from 0.8 to 1.1 solar masses. In his model the mass of the dwarf might then positively correlate with the differences in the absolute magnitude of each event, from –13 to –19. A larger mass leads to a brighter explosion. Low white dwarf mass might trigger a failed detonation or a deflagration meaning that a classic Type Ia event is avoided, while more massive white dwarfs are disrupted in their entirety. Stars with masses in the middle of the range might simply burn up, with the ashes then scattered into space. In these cases much of the star might remain intact.

Intriguingly, modelling suggests that in at least some instances where the underlying white dwarf is not be disrupted the star receives enough of a kick to send it flying off into deep space. In these instances the interior ignites and the erupting flame reaches the surface of the white dwarf producing a jet. This erupts outwards with enough energy to kick the white dwarf through galactic space at speeds up to 11,000 km/h. Precisely why the flame fails in these supernovae is as yet unknown. However, if the white dwarf has too low a mass to sustain the burning front it might generate an explosion that fits observations. Either way the failed eruption produces a seriously underwhelming supernova. SN 2008ha thus represents a deflagration event where the fuel was more limiting or the mass of the white dwarf was sufficiently high that little material escaped.

If SN 2008ha is a failed supernova what might remain will be a bizarre white dwarf with a core rich in iron and intermediate mass elements, such as silicon and calcium.

The truth of these explosions is not yet clear. SN 2005hk is modelled well by a merger between two white dwarfs and an ensuing deflagration involving the whole star. SN 2008hk also works in a variety of models, including core collapse scenarios. The chemistry and behaviour of SN 2002cx implies, however, that these explosions share a common heritage. It seems likely that these weak Type Ia events, Type Iax explosions, may in fact represent the double-detonation of sub-Chandrasekhar white dwarfs as was proposed many decades before. In any case the dye is not yet cast: alternative scenarios may arise that fit the data better. But for now a model, discredited as the cause of generic Type Ia explosions, may have been successfully revived. The double-detonation model may now have found its place as the best descriptor of a unique cause of stellar death.

PTF 09dav: Something Else?

The Palomar Transient Factory has found many of the most unusual and exciting supernovae in recent years. Alongside the wave of ultraluminous events bagged and discussed in Chap. 10, there are many others, equally perplexing but dim by comparison. None, however, is as enigmatic as PTF 09dav, discovered on August 11, 2009.

Like many of the supernovae in this chapter, the paltry mass of ejecta (0.36 solar masses) had a sluggish outward velocity of 6,000 km/s, and the yield of radioactive nickel was, at most, 0.019 solar masses. This dim eruption managed to hit an absolute magnitude of only −15.5, well below even the average core-collapse event, and falling far short of the usual values for thermonuclear supernovae (Table 13.6).

The spectrum revealed the usual suspects, such as sulfur, silicon as well as titanium and calcium, but intriguingly also a unique double bill of strontium and scandium – the former never seen before in significant quantities in a supernova. Strontium is an s-process element with an origin in AGB stars, or more rarely the merger product of low mass white dwarfs – R Corona Borealis stars.

Moreover, the ejecta seemed very cool for a standard supernova, giving the explosion a pronounced red color. The temperature could also be inferred from the amount of ionization of the ejecta. High temperatures would favor the loss of more electrons from atoms. The higher the temperature, the more electrons have sufficient energy to escape their nuclear bonds. In PTF 09dav, many of the elements were poorly ionized, indicating that they were not especially hot when observed. Although the supernova was brief compared to normal Type Ia events, it still took longer to reach its peak than many Type .Ia events (12 vs. 5–10 days).

The problem with this supernova was that nothing quite fitted the bill. The low luminosity could be explained by the deflagration of a helium layer atop a low mass CO white dwarf. Yet the supernova appeared too faint to be explained by the Type .Ia mechanism, so a detonation was probably ruled out. Moreover, both shell ignition models (deflagration or detonation) had trouble producing the observed silicon and magnesium. Yet, a core explosion should produce much more silicon and iron group elements than were observed. The red color, although extreme, is similar to the sub-luminous SN 1991bg-like supernovae. However, despite some qualitative similarities, the low luminosity and very red optical color were hard to explain with a simple model involving a merger of two white dwarf stars.

Moreover, the chemistry of the supernova was certainly enigmatic. The presence of s-process elements, such as strontium, is a novelty for a thermonuclear supernova. Scandium and some others are observed in core-collapse events; these originate in explosive oxygen and silicon burning. Working with others, Mark

Table 13.6 The properties of the peculiar supernova PTF 09dav

SN	Light curve	Ejecta velocity	Ejecta mass	Principle elements	Maximum light	Possible progenitor
PTF 09dav	12 days to peak; an intrinsically red supernova compared to other Type Ia events	6,000 km/s with fairly rapid fall	0.019 M Ni-56 from total ejected mass of 0.36 M	Ca/Ti but Sc and uniquely, Sr; some SI and Mg set this apart from other helium-shell supernovae	−15.5	He-shell detonation or deflagration on surface of low mass CO WD; or perhaps whole star detonation; many contradictory spectral features

Sullivan (University of Oxford, UK) suggested that the strontium originated in the progenitor while it was still a red giant. However, one might expect this in other thermonuclear supernova; after all, the white dwarf progenitor is descended from an AGB star in every case. Instead, what if this element originated in a phase of mixed burning on the white dwarf ahead of the supernova? If the donor star was a post-main sequence object, rather than a white dwarf or helium star, it may have had some residual hydrogen left over mixed into a helium-rich layer lying adjacent to its inert and contracting helium core. A mixed fuel supply could allow some s-processing on the surface of the white dwarf before the subsequent supernova scattered the element during the explosion.

Whatever mechanism caused this element to end up in the debris of the supernova, the combination of elements was certainly singular in nature. Some elements can be explained by a detonation, others by a deflagration. The low luminosity certainly favors the latter, as does the presence of abundant calcium seen in later spectra. Yet, this supernova remains an enigma, awaiting improved modeling and perhaps another example at closer range. Thus far it is certainly one of a kind.

SN 2005E: *Lex Parsimoniae*

Lex parsimoniae is a Latin phrase meaning the principle of parsimony. In science this is usually abbreviated with references to Occam's razor. William Ockham, to whom the expression is attributed, apparently never used it. He did, however, use the principle often, hence the association. Supernova SN 2005E brought *lex parsimoiae* to mind. This supernova, and a close kin SN 2005cz, generated much controversy when analysis of the explosions was published in the journal *Nature* in 2010.[1] Rival groups described these related supernovae variously as core-collapse events or helium-shell conflagrations. To elucidate the probable nature the principle of parsimony – the simplicity of explanation based on evidence – must be applied (Table 13.7).

---

[1] SN 2007cz bears an unfortunate and very similar name to a core-collapse Type II-P event SN 2005cs—Chap. 7. They are not related.)

Table 13.7 SN 2005E-like supernovae. Yet another class of odd supernovae that may be linked to a detonation or deflagration in a helium shell

SN	Light curve	Ejecta velocity	Ejecta mass	Principle elements	Maximum light	Spectroscopic features	Possible progenitor
2005E, 2005ds, 2005cz 2007ke, and SN 2000ds	Slow evolving with low peak but faster decline than sub-luminous Type Ia SN	11,000 km/s	<0.3 M of which nearly half is calcium	Calcium, helium and titanium	−14.8	No S, Si (so Type Ib-like)	He-star donor (larger accreted mass) onto low mass CO WD (0.7 M)

SN 2005E, on the outskirts of the galaxy NGC 1032, was discovered on January 13, 2005, by the KAIT team, led by Alexi Filippeko. Based on the presence of helium in the ejecta the supernova was classified as a Type Ib event.

The supernova was remarkable for a number of distinguishing features. The ejecta was limited in mass (0.3 solar masses at most) and was particularly rich in intermediate-mass elements, particularly calcium and titanium-44. Indeed the richness in calcium was rather overwhelming, accounting for 49 % of the total ejected mass. However, the peak luminosity of the explosion was particularly low, reaching only a few times above that of a typical LBV eruption. Indeed, with an absolute magnitude of only −14.8 it was one of the dimmest supernovae on record.

Two groups proposed radically different mechanism for this event. The Japanese group, led by Koji Kawabata (Kobe University), suggested that this Type Ib supernova, or more accurately a very similar supernova, SN 2005cz, was a core-collapse explosion. They attributed the presence of helium in the spectrum to a typical Type Ib core-collapse explosion.

The rival explanation by Hagai Perets and co-workers followed the shell-detonation scenario for other weak supernovae, suggesting that this was an explosion associated with a white dwarf star. Indeed, the location of the supernovae, far removed from any identifiable site of star formation, favored the latter route. Detonation involved an old, low mass progenitor. Getting a massive star to this location would involve removing the star from the galactic disc and hurtling it outward at speeds that would defy logic. Typical scenarios involving runaway stars just can't produce the necessary velocities. Some form of stellar merger in a cluster might just produce a star of sufficient mass to undergo core collapse, but given the necessary prerequisites for this scenario you are left feeling that it is a little *ad hoc*, and once more Occam (or more precisely William Ockham) can be heard sharpening his razor.

The calcium-rich chemistry of the explosion also pointed to a thermonuclear origin – although this is more debatable. The presence of calcium and titanium was readily explained through nucleosynthesis of these elements from helium fusion under dense conditions.

How can astronomers best explain the particularly intriguing and unique features that set SN 2005E apart from the Type .Ia events and other low-energy thermonuclear explosions? The favored model involves a helium layer – possibly a good fraction of a solar mass in size – undergoing a deflagration. This would produce the very low peak luminosity, the fairly drawn out rise to peak and the faster rate of decline as well as the unusual chemistry of the explosion. In the deflagration model the non-explosive burning led to the formation of the large mass of calcium and other intermediate mass elements but restricted the production of nickel-56 and other iron group elements. The low abundance of these elements, in turn, produced the distinctly underwhelming explosion.

If this all seems rather neat, the velocity of the ejecta was determined to be quite high which, rather contrarily, favored a somewhat energetic event. This combination of features may favor a merger involving two degenerate stars, one rich in helium and the other a more massive carbon-oxygen white dwarf. The low broad curve might imply a progenitor system similar to SN 1991bg-like events, except that in this instance one star is a helium white dwarf, rather than a pair of carbon-oxygen white dwarfs. In this case the lower mass helium white dwarf would be disrupted and form a fairly massive but lower density layer on top of the carbon oxygen dwarf. The helium layer would then ignite, and much of the energy of the resulting explosion would go into disrupting this layer. This would give rise to the broad, low peak – and an explosion rich in unburned helium and intermediate mass elements such as calcium.

What is interesting and a little perplexing is how to generate an explosion that produces nearly half the mass of debris in a single element: calcium. A typical Type Ia event will convert roughly half the mass of the white dwarf to iron and this situation is fairly well understood. However, producing calcium with very high abundance in some form of shell eruption is a little more challenging. The high abundance of a single element implies that the scenario is rather similar to that involving the formation of iron – at least in the sense that a good fraction of the accreted helium must have the precise properties required to synthesize calcium. A more massive, stratified helium layer may be needed with a fairly uniform density, perhaps supporting some sort of merger. The burning front was probably a deflagration rather than a detonation. As it

a. Type Ia Supernova. Chandrasekhar mass CO white dwarf deflagration and detonation. Merger of two sub-Chandrasekhar mass CO white dwarfs or accretion of hydrogen/helium onto sub-Chandrasekhar mass white dwarf.

b. Sub-luminous Type Ia Supernova. Merger of two roughly equal sub-Chandrasekhar mass (~0.9 solar mass) CO white dwarfs; SN 1991bg; this event is super-Chandrasekhar in nature.

c. Calcium-Rich Supernova (Type Ib). Sub-Chandrasekhar mass CO white dwarf with thick/high-mass helium shell SN 2005E. Accretion from He-star or merger of He white dwarf with CO white dwarf?

d. Type .Ia Supernova. Sub-Chandrasekhar mass CO white dwarf with a low mass (<0.1 solar masses) detonating helium shell. Descendants from AM CVn cataclysmic variable binaries?

Fig. 13.4 Possible evolutionary precursor for different kinds of thermonuclear supernovae. The small gradient-filled stars are the CO white dwarf. Blue represents either a star donating helium or, in 15.2a a main sequence or giant donor star. Many of the sub-luminous events may well be super-Chandrasekhar. It is, after all, unlikely that a stellar merger would naturally produce a single Chandrasekhar white dwarf. If so, what then decides how luminous the supernova will be?

progressed through the helium shell, it converted much of it to a single element that was defined by the density and temperature of the layer. The bulk of the helium would have been non-degenerate, or nearly so, and this was converted to calcium. The remainder burned to produce the other observed elements. In this explosion the carbon-oxygen core remained largely untouched.

Modeling of calcium-rich supernovae is in its infancy, but the general picture that emerges so far is that such explosions occur on low mass (0.5–0.7 solar mass) white dwarf stars with a carbon-oxygen composition. Not much else can be said, other than that helium is undoubtedly the fuel (Fig. 13.4).

A detonation might be expected to produce a clutch of heavier elements than was seen and a brighter and faster light curve, perhaps more like a Type .Ia event. However, a larger mass of helium might offset this if the detonation could be avoided and the helium ignited earlier. Where the bulk of the explosion energy is converted into the kinetic energy in the expelled material, the production of radioactive nickel-56 is restricted. This happens where the explosion disperses the material early on or lowers its temperature sufficiently. Instead of iron and related elements, nucleosynthesis produces radionuclides with shorter half-lives, and this drives the observed luminosity. Arnett's rule, relating the outward kinetic energy to the mass of radioactive nickel, is thus somewhat circumvented as other radioactive elements are synthesized in its place. In essence it is the low mass of the underlying carbon-oxygen white dwarf that allows the material to expand and cool as it burns through. The stronger gravitational pull of more massive white dwarfs means that they retain the layer for longer and can advance burning further.

SN 2005E-like supernovae are important biologically, as they appear to be the principle source of calcium in the universe. Combined with a significant yield from Type .Ia events, helium-shell detonations or deflagrations appear to produce the majority of this element. Roughly two such supernovae per century should account for the observed abundance of this element in the nearby universe.

Moreover, these supernovae produce the bulk of the radioactive element titanium-44, which is a significant source of positrons. These, in turn, may contribute a significant fraction of the observed low energy gamma ray emission from the galactic halo. Thus helium-shell supernovae play a very significant role in the physical and chemical evolution of the galaxy. Moreover, much of the calcium in our rocks – and more personally, our bones and teeth – may have originated in these blasts.

Conclusions

This is a peculiar chapter in the sense that this is the collection of oddballs and circus freaks. They appear to unfold in variants of two scenarios: helium-shell detonations or failed carbon-oxygen detonations. These oddities are emerging from the universe of supernovae that are pouring out of the automated supernova searches. An army of professional – and increasingly amateur – astronomers from across the globe are delivering a bewildering array of supernovae that we could only have dreamed possible a decade ago.

What is most interesting about these discoveries? Perhaps it is the unfolding drama of how every element in the Periodic Table came to be. Until the discovery of many of the weird and wonderful explosions described here, the story of elemental manufacture was largely limited to core-collapse and thermonuclear Type Ia supernovae, as well as red giants and cosmic rays. This was much like describing an animal, based on a couple of toe bones or broken teeth. The discovery of calcium-rich supernovae has added more bones to our skeleton, both figuratively and literally, placing them at the heart of the formation of this element.

Prior to the discovery of these calcium factories, a gap existed in our understanding of this element. Now we know from a handful of recent supernovae how the universe synthesizes calcium in the abundance we observe. The organism, as a whole, is coming into perspective.

We should also be considering whether to expand, refine or reconstruct the classification system for supernovae that is creaking under the weight of these new discoveries. Certainly the system is in need of overhaul and simply adding subdivisions to pre-existing classes of event is becoming increasingly unsatisfactory. Improving our understanding of the aetiology of these explosions will in turn produce a richer taxonomy that is also better able to place and link these increasingly disparate events.

So, where are we now? We have two broad groups of supernovae and a third, with perhaps one member. These are the core collapse supernovae, the thermonuclear supernovae and pair-instability supernovae, respectively. The thermonuclear supernovae can be subdivided into at least two broad classes: the core conflagration events and the shell conflagration events. These, in turn, may be split into those involving pure deflagrations and a mixture of deflagration and detonation. The details of these subclasses are not well understood yet. The core-collapse supernovae are currently split into sub-groups according to the identification of different elements in their spectra.

Based on these criteria there are reasonable grounds for splitting the supernovae on mechanistic grounds: core collapse; thermonuclear core or whole star; pair instability and shell thermonuclear events. These last two perhaps could even be regarded as types of novae, given the strong similarity in their underlying mechanism. At the time of writing with only two, or at a stretch three members, the pair-instability supernovae are currently on the shakiest ground.

Looking specifically at the thermonuclear supernovae, there is now substantial evidence in favor of the double-degenerate model. Both peculiar supernovae such as SN 1991bg (Chap. 13) and more typical supernovae like SN 2011fe (Chap. 11) and the historical SNR 0509-67.5 are best explained by models involving the merger of white dwarf stars. There is some evidence for a hydrogen-rich donor star in SN 2011fe, but it must have been relatively small not to have contributed to the light curve of the supernovae, or to have appeared in archival images of the region in which the supernovae erupted. A combination of effective and extensive archiving of galactic images has thus constrained the nature of the progenitors in a growing number of core collapse and thermonuclear events. In turn this both constrains and challenges existing paradigms regarding the nature of the mechanisms of these supernovae. Exciting times, indeed.

Part IV
In Flagrante Delicto

14. The Mysterious Case of V838 Monocerotis and the Red Novae

Introduction

Not all stellar explosions are associated with the death of stars. Astronomers are familiar with novae, the explosive but non-fatal eruptions associated with white dwarf stars in binary systems (Chap. 13). They are also increasingly aware of temporary eruptions in very massive supergiant and LBV stars (Chaps. 3 and 8).

However, it is also apparent that there are other sorts of eruptions that are caused by the merger of stars. The vast majority of stars are born in clusters, even those which, late in their lives, travel the galaxy alone. In such tight-knit groupings stars can encounter one another and occasionally collide. In Chap. 6 we saw that Simon Portegies Zwart had suggested that the outburst of SN 2006gy was caused by a stellar merger. But what of lower mass stars? What happens when two lower mass stars collide? Have the fireworks of these sorts of random, but possibly common, events been observed? Would astronomers know what to expect in such situations? In recent decades a few odd eruptions have been witnessed that might just be the outcome of merger events. Such "mergebursts" are becoming a recognized explanation for some eruptions – the products of which may be stars such as the blue-stragglers commonly found in star clusters (Fig. 14.1).

324   Extreme Explosions

Fig. 14.1 The V838 Eruption of 2002 showing an initial blue explosion, revealed by the outer blue light echo in the May 2002 frame followed by the deep red phase of the main eruption (Image courtesy of NASA, HST)

V838 Moncerotis

In January 2002 V838 Monocerotis achieved what few other stars have managed over the last century – front page headlines across the globe. An unassuming star had brightened dramatically until it outshone the majority of its siblings in our galaxy. Reaching an absolute magnitude of −9.6, it rivaled the great eruptions of luminous blue variables (LBV).

Over the ensuing weeks and months the Hubble Space Telescope took several panoramic shots of the star and its surroundings, revealing a brilliant red supergiant embedded within extended reflective nebulosity. The earliest photographs revealed blue echoes reflecting off these dusty clouds. This suggested that despite the later red color, the initial explosion had been distinctly blue in color.

Over time the star continued to cool until by May 2002 it had become the prototype of a new kind of star, an L-class supergiant. L-class objects had previously been confined to the dim, bottom right-hand corner of the HR diagram, the preserve of the coolest red dwarfs and warmest brown dwarf stars. Numerous stars and brown dwarfs of this class had been detected by the infrared 2MASS

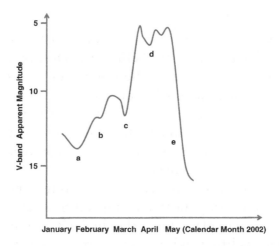

Fig. 14.2 The light curve of V838 Mon during its 4 month long eruption. (a) Dip in the green (v) band output prior to the outburst; (b) Initial rise in January associated with the blue phase of the eruption; (c) Dip prior to "peak" of the eruption; (d) Broad plateau with three smaller peaks associated with red maximum outburst; (e) Steady and rapid decline through the late spring and early summer of 2002

project in the late 1990s, but no evolved giant or supergiant star belonged to this class.

The light curve of this explosion was also unique. It bore several distinct peaks, the first blue, followed by a much brighter red summit containing another two (Fig. 14.2). This intricate arrangement of peaks and troughs indicated that a very complex event had occurred. Was this some sort of nova, an outburst of an LBV, or something completely different?

The first clues to the nature of this event came from an investigation of the stellar environment and from a fortuitous spectrum taken of the progenitor before the explosion. These observations, coupled to detailed analyses of spectra taken throughout the eruption provided astronomers with many clues and led to a variety of explanations for the blast. Ultimately, only one explanation really fitted the bill, and it was unique. However, before examining the winning explanation we look at the other scenarios in turn.

In the first, the star that underwent the eruption was highly evolved. It has already left the main sequence and had subsequently consumed the bulk of the helium ash produced throughout its life.

This was an aged thermally pulsating Asymptotic Giant Branch (AGB) star. As we saw in Chap. 7 stars with low to intermediate masses evolve through this AGB phase. During this time they alternate waves of hydrogen and helium fusion in thin shells. Initially, hydrogen burns adding helium ash to a layer atop the inert carbon-oxygen core. During this time the star is very bright and draws almost all of its luminosity from hydrogen fusion. Once the helium layer has grown massive and hot enough to burn, it ignites explosively under degenerate conditions.

Helium fusion is extremely sensitive to temperature, and as the degenerate layer traps heat, its temperature soars and explosive helium burning takes place. Once the shell has become sufficiently hot, degeneracy is lifted and the layer expands upwards. Although not observed, the star is expected to initially brighten dramatically as the heat from the helium shell bolsters fusion in the hydrogen burning shell above. However, as the helium layer expands the hydrogen burning shell is shoved upwards, cools and fusion largely or completely shuts down. The star then dims and contracts until the helium becomes exhausted and hydrogen fusion can recommence. This process repeats on timescales of thousands of years until the hydrogen-rich envelope is expelled as a planetary nebula. Was the outburst of V838 Mon a helium-shell flash?

Several observations cast doubt on this interpretation. Clearly, if V838 Mon was an AGB star it would be highly evolved, and the spectra of the star would show both the presence of abundant carbon or oxygen produced by helium fusion, as well as so-called s-process elements, such as zirconium or technetium. Neither type of element was seen. Moreover, V838 Mon was clearly associated with a cluster of hot, young and fairly massive blue B-class stars, indicating that the eruption occurred within a young star. Finally, fortuitous observations of the region prior to the outburst indicated that the progenitor of the explosion was also a young B-class main sequence star (B3V), with around 8 times the mass of the Sun. Clearly, it was not something that had already left the main sequence.

Another scenario involved a nova-like explosion on a white dwarf star. Once again, the youth of the star cluster to which V838 Mon belonged refuted such a scenario. An 8 solar mass star would take around 20 Ma or so to reach its white dwarf finale. This star was a lot younger, even than this.

An intriguing third scenario involved the consumption of planetary companions by a dying red giant star. In this case, as the red giant expanded it swallowed a succession of massive planetary companions. Each time one was consumed it added that planet's kinetic energy to the envelope of the giant, which caused it to inflate further, which in turn allowed it to ingest another planet lying further afield. It was suggested that these giant planets would then rapidly spiral inward and add fuel to the hydrogen-burning shell at the base of the envelope. Each initial expansion brought the expanding star another planetary meal, and so the process continued until all three of its planetary companions was eaten. Each companion triggered a peak in the light curve of the explosion. However, the scenario seemed rather contrived. Just how quickly could a planet be absorbed by the star, and what would the pattern of brightening look like with each meal? Moreover, it was certainly unclear whether such a massive star could produce planets in the time it had resided on the main sequence.

The age of the system called all of these scenarios into question. V838 Mon is associated with a large mass (perhaps as much as 900 solar masses) of dusty hydrogen and helium. There were suggestions that this material had been shed by V838 Mon in previous eruptions. However, the sheer mass of this cloud made it clear that this interpretation was untenable. Instead it was assumed that this dusty cloud represented the remains of the birthing chamber that spawned V838 Mon and its associated cluster of B-class stars. This, too, implied a young age for the star system and for the progenitor of the explosion. Thus all explanations involving the late stages of stellar life were rejected. The eruption of V838 Mon had to be caused by something else.

If we abandon each of these explanations, what is left? Perhaps the most promising candidate is some form of stellar merger. The "mergeburst" model was proposed by Noam Soaker and Romuald Tylenda, and although not without criticism, it is by far the most successful of the models thus far proposed to explain the V838 Mon eruption.

Very little is understood about how the drama of stellar mergers would unfold and, more pertinently how these events would appear to observers in the outside universe. However, it is clear that many stellar clusters possess unusually massive stars that appear to defy the logic of stellar evolution. These blue stragglers are so-called because they appear to be hanging onto their main sequence existence well beyond their sell-by dates. They sit uncomfortably above the turnoff point where the vast majority of main sequence stars give up the ghost and retire to the pavilion. In denser star clusters it is thought that most of these blue stragglers emerge through the process of stellar collision, achieving higher than expected masses by merging with neighboring stars. In lower mass and less dense clusters blue stragglers appear to be formed primarily through accretion of matter of one star from another.

However, the open star cluster associated with V838 Mon is hardly a bustling metropolis. Given its more suburban location, could a stellar merger explain the eruption? Let's re-examine V838 Mon this time up close and personal.

Pre-explosion spectra suggested that the progenitor was part of a close-knit binary system, with both components bearing fairly similar B3V spectral classifications. During the eruption the brighter of the two stars appeared to have been the culprit, expanding rapidly and cooling as a result. At some point in the eruption, the less luminous companion was engulfed. Later spectra suggested that some of the supergiant's material was accreted through a temporary disc onto the dimmer companion. Finally, once the outburst was complete, the now red supergiant, primary star contracted once more, and the material around both stars began to disperse.

So, if the companion survived the eruption, what was it that triggered it? The presumption was that V838 Mon was actually a triple-star system, with an unseen, third and lower mass star. The system was young and still settling down. Such triple systems can be unstable, and in the mergeburst model it was this instability that was the source of the final collision and merger.

Imagine that prior to the eruption, the V838 Mon system consisted of two relatively massive stars in a wide and eccentric orbit. Around the more massive star, and much closer in, was a third low mass star. As the two massive stars adjusted their mutual orbits around their center of mass, they began to exchange angular momentum with the third star. The smaller star started spiraling inward as the orbit of the outer star increased and became more circular. Eventually, the small star spiraled into the envelope of

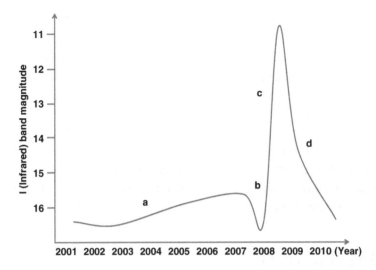

FIG. 14.3 The light curve of V1309 Scorpii showing initial brightening before a sudden brief dimming and subsequent rise to maximum light. This is followed by a rapid decline. (a) Gradual increase in luminosity caused by tidal heating and formation of initial disc of hot material. (b) Rapid dimming as the excretion disc thickens and the merger commences. The disc blocks light from the merging stars. (c) Rapid brightening as the merged stars expand rapidly, dissipating energy from the merger event. (d) Contraction of the newly formed star causes final dimming

the B3 primary star. During the initial part of the death-dive, the exchange of angular momentum drove material outward from the larger of the two stars involved in the collision. This formed an expanding *excretion* disc around the merging pair of stars (Fig. 14.3). Once the merger was complete, the disc stopped forming, and a torus of material was created around the merger product. Interestingly, it has been suggested that this torus of material could be a likely place to form planets, and thus the death of one star may in turn be associated with the birth of planets.

The large kinetic energy associated with the collision then inflated and drove off part of the envelope of the more massive star, while the denser, lower mass star continued to plow inward, eventually merging with and becoming disrupted by the core of the more massive primary. The first wave of light was caused by the initial collision, while the later, rapid rise in luminosity to the extended peak was caused by the collisional heating of both the envelope and the core.

What next would we expect to observe if the mergeburst model was correct? Over the next few years the primary star should contract and begin to heat up once more. Presumably there should also be detectable changes in the orbital period of the surviving B3 companion star in response to the merging of the two innermost stars.

Although far from proven, the mergeburst model works well, and the scenario proposed for V838 Mon is no longer a model without precedent or antecedent. Two other red novae, one recent and the other predating the eruption of V838 Mon, now add weight to the mergeburst model.

V4332 Sagittarii

In 1994 V4332 Sagittarii suffered an outburst that was qualitatively similar to V838 Mon. A previously unassuming star became far more luminous and cooler, indicating that it had expanded greatly over the course of a few weeks. Initially observed as a warm, orange K-type giant with a surface temperature of 3,500–4,500 K, it cooled until its spectrum resembled that of an M5 or M6 supergiant object, with a surface temperature of around 2,500 K. In 2009 T. Kaminski, R. Schmidt and Romuald Tylenda investigated it using Chile's Subaru telescope. They found some interesting features that were best explained by an outburst caused by a stellar merger.

Spectra taken of the system found that the central star was now a large red giant or supergiant star surrounded by a thick, opaque disc of dusty material. Observations suggested that astronomers were observing the disc edge-on. Meanwhile, emission lines indicated two bipolar outflows emerging from above and below this disc of material.

This situation is very reminiscent of highly evolved giant (AGB) and supergiant stars such as IRAS 18059–3211, known as Gomez's Hamburger. Other spectral features resemble those seen in supergiants such as VY Canis Majoris. Thus V4332 Sagittarii had transformed itself into a luminous but very cool supergiant star over the course of a few weeks in 1994. Clearly, standard stellar evolutionary scenarios cannot account for this transformation, and a more violent phenomenon must have been responsible.

Although the underlying cause of the event was unknown there existed a clear synergy with V838 Mon, which in turn suggested a common mechanism.

A Stellar Merger Caught in the Act

If V4332 Sagittarii was an event missed, while V838 Mon teased and suggested, a third eruption in the constellation of Scorpius proved to be more fortuitous for astronomers. On September 2, 2008, a burst happened for which there were good observations of the system before, as well as during and after the event. The eruption of V1309 Scorpii was caught *in flagrante delicto*.

The system giving rise to this eruption was observed as part of the OGLE project (the Optical Gravitational Lensing Experiment, based at the University of Warsaw) from August 2001. There was clear photometric evidence that the "star" was actually a very close pair of stars orbiting one another every 1.4 days. Light from the system showed small up and down variations in brightness as one star passed in front of the other, periodically blocking light from it. Measurements also showed that the orbital period was decreasing with time, indicating that the two stars were closing in on one another and would soon merge. In 2008 that process climaxed, and the two stars became one.

The first spectra taken of the rapidly evolving, post-merger object was of an F-class (7,500 K, yellow-white) giant star. Over the next month this "star" progressively cooled until its surface bottomed out at 3,500 K. It now resembled an M-class giant. Seven months later it had cooled to around 2,500 K. The spectra became rich in absorption lines typical of these "late" giant stars. However, like V4332 Sagittarii, spectral emission lines indicated broad outflows with velocities ranging from 150 to 1,000 km/s – considerably faster than most winds from giant stars. Superimposed upon these lines were shallower absorption dips resembling P-Cygni lines seen in supernovae. These suggested that the outflows were asymmetrical, with material streaming most rapidly away from the polar regions and at a more leisurely pace in the equatorial band. This waist clearly resembled the donut of material inferred from the spectra of the V4332 Sagittarii system.

332 Extreme Explosions

The rate of expansion of the fastest-moving material was far slower than seen in novae. Moreover, the very red spectra also ruled out this interpretation.

Spectra and photometric observations were crucial in determining what was going on throughout the eruption. Pre-eruption spectra suggested that the primary star was the source of the outburst and that it had the spectrum corresponding to the orange K-class giant star. Data suggested that this was a star with roughly the mass of the Sun that was evolving into a red giant. In such a system, as the primary star expands, if the mass ratio with its companion is just right, this secondary companion star raises tides on the surface of the giant primary star. This slows down the orbital speed of the secondary, causing it to spiral into the giant primary star. Such a process is expected to doom Earth in 5–6 billion years when our star evolves into a red giant.

Photometric observations showed that the star was dimming in the month prior to the outburst, indicating that the interaction between the two stars had begun and a disc of material was being excreted along the equatorial axis. This would be expected if angular momentum was conserved. The two stars move together, and a mass of material moves outward to compensate for the change in the momentum of the two stars. At even earlier times, back to 2002, the system appears to have been slowly brightening as orbital energy was transferred into the primary star. Once the stars were close enough, the primary began to shed material and the disc formed, dimming the stars. Finally, in January 2008, the stars begin their final dance and start to merge. Through February 2008 the periodic dips and brightening in the light curve disappeared, indicating that the merger was underway.

Once the merger had commenced there was a steady change in the light curve, with a doubling of the brightness every 19 days until August 2008. During this phase the secondary star was spiraling inward, transferring its orbital kinetic energy to the envelope of the star. This was the source of the energy for expansion. Finally, during August 2008, the rate of brightening accelerated by a factor of 300 over 10 days. It is likely at this point that the dense secondary star had finally transited the entire envelope and reached the dense core of the giant. Delivering a hefty payload of hydrogen to the hydrogen-burning shell, the energy output of this layer dramatically increased. This caused the star to brighten spectacularly.

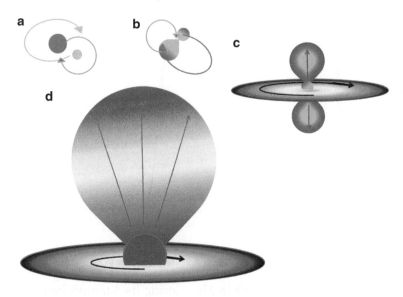

FIG. 14.4 Anatomy of a mergeburst. In (a) and (b) two stars gradually spiral together, and material is then shed from the larger star into orbit around both. At (c) the two stars have merged, forming a very hot object surrounded by an expanding disc of cooler material. At (d) broad outflows have been generated, and the central star continues to expand and cool. After some time, possibly decades or centuries, the star will contract and heat up again forming a blue straggler

During the merger hydrogen would also be dragged into the hot, helium core, causing potentially explosive burning. This would also add to the luminosity of the star.

The eruption was partly damped off by the envelope of the giant. Some of the angular momentum of the two progenitor stars was shed from the system through broad outflows. These escaped along the rotation axis of the star. Remaining energy was restricted in its escape in part by the thick torus of material around the stellar equator, driving a slower outflow in this direction.

During the subsequent collision the orbital kinetic energy of the denser secondary star is transferred to the primary giant star envelope causing it to become hotter. The extra energy, therefore, changed the spectrum of the primary star to a hotter, F-class giant. The surge in energy also caused the star to inflate and become much brighter. The swollen star eventually became 10,000 times more luminous than it was before, radiating 10^{37} J of energy per second (Figs. 14.4 and 14.5).

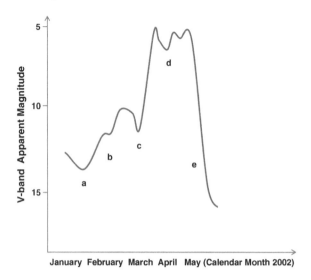

FIG. 14.5 The light curve of V838 Mon during its 4-month long eruption interpreted through comparison with V1309 Scorpii. (a) Dip in the green (v) band output. Is this the formation of an excretion disc? (b) Initial rise associated with the initial merger; (c) Dip prior to "peak" of the eruption, star partly hidden by disc; (d) Broad plateau with three smaller peaks associated with collision of stellar core with smaller denser star and release of orbital energy in the stellar envelope; (e) Steady and rapid decline through the late spring and early summer of 2002

The final display of V1309 Scorpii was caught by OGLE. Once the eruption had peaked, OGLE detected a relatively swift decline in the light emitted from the system. As the central star began to contract and heat up once more, it began to flicker. The light varied erratically as the shrinking central star began to shrivel behind the wall of debris excreted during the eruption.

At some point in the future, the dusty torus of debris will be cleared by stellar winds from the giant star, and the object will finally brighten once more. It will then achieve its final luminosity as a more massive red giant, perhaps surrounded by a retinue of young worlds condensing from the torus of material shed during the eruption. In terms of its altering the evolution of the giant star, the collision may have had less impact than you'd expect. The extra mass in the hydrogen-burning shell and the envelope certainly made the star initially brighter than it would otherwise have been. However, it is the mass of the helium core that has the greatest impact on the fate of a star. Adding hydrogen to the hydrogen-burning shell may make the final mass of the helium core somewhat higher than it would otherwise have been. However, the majority of the accreted mass will simply be dispersed by stellar winds, leaving the giant pretty much as it would have been before the merger. Thus, dramatic though the merger was, the companion's toll may simply be a temporary blip in the life of the primary star. This may be a dog with lots of bark but little bite.

M85's Red Nova

In one final interesting example a red nova was observed in the dwarf spheroidal (SO) galaxy, M85, in January 2006. Known by the unglamorous name of M85 OT20061 this peaked at an absolute magnitude of −12, approximating with the brighter end of LBV eruptions but just short of most (or perhaps all) core-collapse supernovae.

Two very different explanations were published in the journal *Nature* in 2006. Andrea Pastorello interpreted this event as a faint core-collapse supernova − a peculiar Type IIP explosion − while Shrinivas Kulkarni (Caltech) and colleagues propose that this was a mergeburst event. In support of the former interpretation, the explosion was brighter than other mergebursts hitherto observed. However, a supernova explanation demanded some support. Pre-explosion images were used to rule out the presence of a massive progenitor star, though they were insufficiently sensitive to exclude stars with masses in the range 8−9 times that of the Sun. In order to explain the lack of strength of the explosion, Pastorello's supernova ejects less than a solar mass of material and produced very little radioactive nickel-56, with a total energy of less than 10^{43} J. However, even with the likely electron-capture route, something Pastorello and others had suggested for SN 2008S, it really wasn't clear why the mass of ejected material was so low.

Kulkani and co-workers attacked the core-collapse scenario, drawing attention to the lack of evidence for other more massive stars or the presence of active star formation. Instead they showed that the red color of the event best matched the mergeburst hypothesis. The only problem was the larger than expected luminosity for such an event.

The merger of an eight solar mass star with a smaller sibling – as was touted for V838 Mon – produced a far less luminous explosion than the M85 transient by an order of magnitude. The closet analog described by Kulkani and colleagues was M31's red variable (RV). However, a merger model for the latter has its problems. The light from this object is now very blue, in contradiction of expectations from both models and from observational evidence of other merger events.

Therefore, the bright M85 optical transient may have another, distinctive cause, unrelated to either core collapse or stellar merger. We often forget that relatively little is known about the details of stellar evolution. Many of the broad brushstrokes are clear, but a significant number of the transitory phases, where stars actively evolve from one style of living to another, are poorly understood and have been rarely observed. After all, these changes may be very brief and thus are extremely difficult to catch in the act. It is to the act of undressing that we turn to next in the penultimate example of seemingly inexplicable behavioral change.

The Case of the Optical Transient NGC 300 OT

"Opisthodromism" is a marvelous, fabricated word. Coined by Icko Iben Jnr and related by Robert Rood in the 1996 conference proceedings *Advances in Stellar Evolution*, it effectively means being dragged backwards while running forwards. In the context of stellar evolution, many evolved stars track back towards the hotter, bluer part of the HR diagram while evolving away as red giants.

In 2008 a bright transient was observed in the spiral galaxy NGC 300. The object was distinctly blue at first, releasing 10^{40} J of energy over the course of the eruption, with a peak absolute magnitude of −12. This is comparable to an outburst from a luminous blue variable (LBV) and to the M85 optical transient discussed above. However, upon examination of archival infrared images, Howard Bond found no evidence that something as brilliant as an LBV was present. Instead, there was clear evidence for a bright but heavily dust-obscured object with a luminosity 55,000 times that of the Sun. The stellar output suggested a star with a mass 10–15 times that of the Sun, and it was this that had undergone some form of outburst.

Intriguingly there was a clear similarity between this transient and the "supernova imposter," SN 2008S. Both objects were initially buried in copious amounts of dust; both had luminosities tens of thousands of times that of the Sun; and their inferred masses were comparable. Unlike SN 2008S, the Optical Transient (OT) in NGC 300 provided ample clues to what was taking place. There was no suggestion that this was a supernova – or an outburst of an LBV.

NGC 300 OT, the name under which this object gloriously languishes, appears to be an example of opisthodromism, the rapid transition from a red to a blue supergiant phase. This is a star that is seemingly moving back toward the main sequence from the red supergiant branch.

After the initial outburst the OT spectrum reddened somewhat with time, in a manner similar to V838 Mon. There was evidence for the development of a strong, bipolar stellar wind. Before the outburst the OT was clearly heavily obscured and in the red supergiant phase. The star was buried in an oxygen-enriched shroud of dusty material, probably originating from the star itself. However, a few hundred years or so after burying itself in its own dust, the star appeared to have instigated a new phase of its evolution. It was now busily shedding this grimy cloak to reveal a hotter, yellower, fresher F-class star.

In evolutionary terms this may be the first observation of a star instigating a so-called blue loop. The star leaves the red giant branch and returns to the hotter, bluer regions of the HR diagram, possibly as a result of helium ignition in its core. The timescale for these sorts of changes are unknown, but are likely very brief. In massive stars helium ignites gently and progressively builds in intensity until it provides the bulk of energy production in these giants. Stars become bluer before settling in a region just above the main sequence. Most intermediate and massive stars are expected to undergo this change during their post-main sequence evolution. NGC 300 OT's outburst may be the first observation of such a transition. In this regard, the supernova imposter SN 2008S may be another example of such an evolutionary change, rather than a supernova. It is sheer coincidence that both were caught in the same year.

Once Red, Now Blue

A final and far more ambiguous eruption occurred in M31 in 1988. M31-RV achieved a peak absolute magnitude of −9.95, rather less than either the NGC 300 OT or SN 2008S, but comparable with V838 Mon. At the time it was suggested that this was yet another example of a mergeburst – an object undergoing an outburst following a collision or merger event. However, spectra taken subsequently show a very hot, UV bright object at the location of the outburst.

As we have seen, a mergeburst should produce a cool but slowly contracting and warming star with a very distinctive spectrum. M31-RV is currently far too blue to have resulted from a mergeburst, unless somehow it has rapidly shed its cool, outer layers. Instead it is generally assumed that the most likely explanation for this eruption was a nova, a thermonuclear explosion on the surface of a white dwarf star. The only real problem with this scenario was the peak brightness of the event. More than most novae but dimmer than a supernova. Is the host of the M31-RV the survivor of a shell detonation event? The star resides in the stellar bulge of the galaxy, implying that it is a rather old star. Other than that little else is known about it. Although neither the nova nor mergeburst models currently fits the bill other explanations may arise through further observation, and it reminds us that there is still much to learn about the lives and deaths of stars.

What Do These Collisions Tell Us About Stellar Eruptions?

Stellar collisions have the propensity to produce powerful displays rivaling luminous blue variables. These explosions are unique in that they evolve over several weeks and months and are very red in color. The outcome may be a single, more massive star that may have a fundamentally altered evolution compared to its progenitors. Most significantly, a large body of work has focused on the final state of such merged objects, but until V8383 Mon and its contemporaries were observed, little was known about how such merger events would unfold. The eruption of V838 Mon opened a window on a very dramatic and brief phase of stellar evolution that had been hitherto unobserved.

The Great Eruption of Eta Carinae: A Reprise

The 1843 eruption of Eta Carinae was almost without exception. At peak, the apparent magnitude of the event (as it appeared to the naked eye) exceeded Sirius, Canopus and Alpha Centauri, making it the brightest stellar object of its time. This was despite the great distance to the star. As investigations by Nathan Smith and others have shown, the eruption released almost as much energy as a supernova and launched several solar masses of material into space at thousands of kilometers per second.

Today, Eta Carinae is flanked by an expanding equatorial disc of material and vast, burgeoning, bipolar lobes of dusty hot gas. The star is embedded in material that is so dense, it is effectively no longer star-shaped. Eta Carinae's bipolar outflows have instead turned the star into an enormous, luminous peanut. The stellar surface, its photosphere, is the region where stellar density drops sufficiently for photons of visible light to freely escape. However, the bipolar outflows are so dense that light escaping the star's polar regions is trapped until it is a considerable distance from the center of the star, much further than at the stellar equator. This distorts the effective surface of the star from ellipsoidal to that of a peanut. Given our observations of mergebursts can the Great Eruption be understood, not as the outcome from some vast internal machination but rather as a result of a collision, or merger, event in the Eta Carinae system?

The vast majority of massive stars appear to reside in close binary systems. Modeling suggests that at least 10 % of these systems will see their massive stars merge during their lifetimes.

It is, therefore, not improbable that mergeburst events will be common in systems containing massive stars. So, could the presently eccentric, massive binary Eta Carinae system be a descendent of an unstable triple-star system, much like V838 Mon?

In Chap. 3 we explored scenarios involving unstable nuclear burning or energy transport. Here, we examine the mergeburst model for the Great Eruption. There are a few interesting pieces of evidence in this scenario's support. First, the energies involved are comparable to what would be expected in a collision between this very massive star and a less but still formidable companion. Second, the structure of the outflows is very similar to that observed at V1309 Scorpii, V4332 Sagitarrii and V838 Mon – a dense equatorial torus punctuated by two lobes of gas expanding from the polar regions of the star. Finally, the Eta Carinae system appears to be a binary like V838 Mon. Periodic swings in the X-ray output of the system point to the presence of an unseen but massive companion in a 5.5 year orbit around the system's center of gravity. As this secondary star swings in close to Eta Carinae, colliding stellar winds generate an upswing in the amount of X-rays emitted by the system. The shape of the orbit is similar to that of the B3V companion to V838 Mon.

In the late 1990s Icko Iben Junior suggested that the great eruption was caused by the collision of a companion with Eta Carinae.[1] Subsequent work by Philipp Podsiadlowski modeled the outcome of such a merger event. He begins with two stars in close proximity and a third more distant star, the current 30 solar mass companion invoked to stimulate the swings in the X-ray emission of the system. As the system settled with age, the orbit of the closer of the two companions became unstable. This may have been in response to gradual alterations in the orbit of the outer companion star, or because the largest of the three stars was evolving and expanding toward it. At some critical point tidal forces between the doomed companion and Eta Carinae's already unstable envelope caused the fated star to spiral into Eta Carinae. Release of gravitational potential energy and the orbital kinetic energy of this

---

[1]"Eta Carinae at the millennium." ASP Conference Series, vol. 179 (1999). J.A. Morse, R. M. Humphreys and A Damineli eds. p373.

star heated and expanded the envelope of Eta Carinae, causing it to brighten, as was observed in 1843. Finally, Podsiadlowski's calculations suggested that as the core of the smaller, denser companion reached the base of the hydrogen envelope of Eta Carinae, a substantial amount of hydrogen was mixed into the very hot helium core of the star. This triggered explosive nuclear burning, and the star violently expanded and brightened in response.

Eta Carinae is so massive that it was likely already on the edge when the merger occurred. The current star emits enough radiation to risk blowing away its entire stellar envelope. The star resides close to the so-called Eddington limit. Adding extra fuel to the fire tipped the balance, resulting in the violent ejection of several solar masses of gas in the Great Eruption. However, measurements of the velocities of the ejected material by Nathan Smith put some of them at over 6,000 km per second, comparable with those seen in supernovae. Whether a mergeburst can generate such high velocities – even with a star as massive as Eta Carinae – is anyone's guess at present.

We will probably never be certain what caused the Great Eruption of 1843, but a mergeburst scenario certainly explains some of the peculiar features of this event, which are less readily explained by pulsational pair instability or an instability within the stellar envelope.

Is the mergeburst model without precedent in massive stars? Well, quite apart from the predicted merger rate, a number of other massive stellar objects have properties that might best be explained by stellar merger events. SN 1987A erupted within a well characterized but mysterious hourglass shell of gas. The very massive star Sher 25 in NGC 3603 has a virtually identical structure around it, while a few other massive stars are flanked by similar structures. Although it is not unusual for stars to emit winds in specific, localized directions, to form such elaborate structures seems to require the influence of a companion star (or planets) in the case of more abundant planetary nebulae.

In the case of SN 1987A it is suggested that the red or blue supergiant that would ultimately give rise to the supernova underwent a merger with a companion 20,000 years before the supernova occurred. The companion star spiraled into Sanduleak −69° 202, ejecting the bipolar and equatorial rings of gas and dust that are currently being smashed by the supernova. In the process, the condensed stellar envelope was infused with helium from the stellar core and contracted, heating up, until the formerly red supergiant became blue.

Once again this is a scenario, a model, that can reproduce the conditions that are observed. It is not proof that this occurred, but the scenario does recreate the circumstances in which the supernova happened and, therefore, warrants further investigation. Whatever the outcome, it now seems certain that mergebursts will become an increasingly important feature of stellar evolutionary models.

Conclusions

Mergebursts, as these collisional events are now widely known, are an increasingly widely applied explanation to a number of unusual and unexpected stellar eruptions. It had been known for some time that stellar mergers must occur, but the frequency of such events was probably underestimated. Increasingly detailed observations of stellar clusters, plus an improved understanding of the events that accompany the merger process, are underlining the significance of merging in altering the fates of stars in clusters and in higher-order systems with three or more stars. It is now apparent that the collisions between stars may be very dramatic indeed. Moreover, there may be spectacular consequences for the surviving star following the collision. This may be all the more true for massive stars where the kinetic energies involved in the collision will be greater and for which the stellar core manufactures more energy. For one thing, it seems likely that the very massive stars produced by collisions may have radically different fates than originally thought (Chap. 6). Further modeling of both the probability of collisions and the outcomes of these is clearly warranted.

15. Between Scylla and Charybdis

Introduction

The taxonomic scheme devised in the 1950s has held up remarkably well, considering the dramatic growth in supernovae detection. In 1998 20 supernovae were discovered, but by 2007 this figure had risen to over 500 per year. With a vast array of automated medium-sized 'scopes' and a sizable army of dedicated astronomers, supernovae are pouring out of the most advanced search engines at the rate of nearly 1,000 per year.

Perhaps it is remarkable that the vast majority of the supernovae detected in the last decade fall neatly into the pre-existing categories. One might casually expect large numbers of extravagant or peculiar events to turn up. However, most are run-of-the-mill explosions for which astronomers feel they have *at least* a modest grasp of understanding. Yet among the populace of events are a few belligerent and precocious objects that defy casual inspection and classification. It is these supernovae that demand a better system of classification. These are the diamonds in the rough.

Type IIa?

Where is the system breaking? A casual example is the introduction of Type IIa supernovae: thermonuclear events that show interaction with circumstellar hydrogen and thus display characteristics of both Type Ia and Type II events. OK, maybe this is another acceptable rarity that we can simply sweep under the rug. Yet, we also have core-collapse supernovae that initially show hydrogen but later lose this spectral feature. These are called Type IIb supernovae.

One might expect Type IIa to be related to Type IIb, given the logical development of their names, yet obviously they have

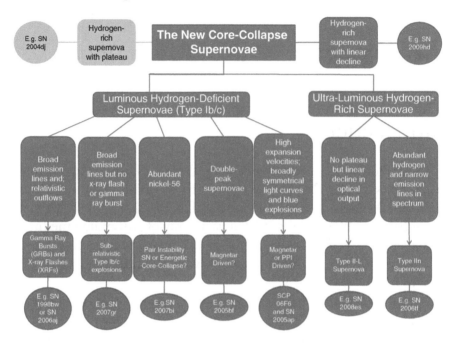

FIG. 15.1 The core-collapse supernovae. The taxonomic tree includes all of those types identified at the time of writing – although I wouldn't dare suggest that this is complete. The stereotypical original class is shown in *yellow*

nothing in common. Moreover, the mode through which hydrogen becomes evident in the spectrum couldn't be more different. The logical flow is lost.

Then we have the so-called hypernovae, or broad-line Type Ic events and their more conservative cousins, the narrow-line Type Ic explosions. These at least show a chemical and probable evolutionary kinship. However, the introduction of the whimsical Type .Ia explosions is pushing the limits of an acceptable classification system. You won't find many biologists classifying a small, new primate as a ".Hominid." The sub-luminous, SN 1991bg-like, Type Ia explosions could do with separation from the others, as their chemistry and mechanism seem distinct. The system is in need of a thorough review and almost certainly an extensive overhaul.

As the two classification trees (Figs. 15.1 and 15.2) indicate there are many new forms of supernovae that could do with a reasonable place in the system. It is notable that astronomers don't

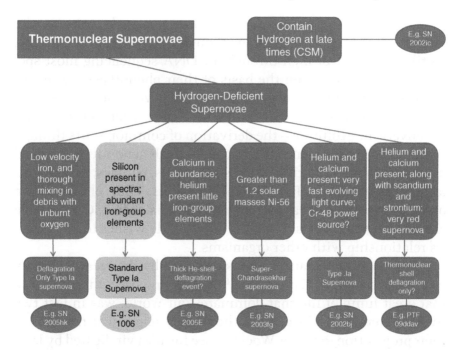

FIG. 15.2 The current taxonomic scheme for the thermonuclear supernovae. Once again, the stereotypical, original class is shown in *yellow*. The Type .Ia supernovae and the calcium-rich supernovae might be grouped into a third, distinct group called *shell supernovae* – and may then sit more comfortably with the novae, with which they share a similar underlying mechanism

have a clear mechanism with which to approach the problem of some of the thermonuclear supernovae, or the luminous blue flashes. Indeed, one could reasonably place SN 2007bi in with the thermonuclear supernovae that have kinship with Type Ia events. After all, the underlying mechanism are related (explosive carbon burning), even if their causes are distinct.

Much of the classification of new and unusual supernovae is hearsay and guesswork, based on a few scattered clues. How might we identify and classify rare supernovae caused by the tidal disruption of stars by supermassive black holes or resulting from the collision of white dwarf or neutron stars with main sequence stars? They certainly aren't core collapse or thermonuclear in the traditional sense. Many astronomers, such as Philipp Podsiadlowski, are already questioning the current classification scheme.

The current situation in astronomy closely mimics the revolution that DNA sequencing brought to biological phylogeny. Until the advent of high-throughput DNA sequencing most species were understood on the basis of their phenotypes – a mirror for spectra in the world of supernovae. However, phenotypes are a messy business, with some features arrived at through the process of convergent evolution – the derivation of common traits through a common environment rather than through genomic (genotypic) descent through mutation. This is analogous, perhaps, to the similarities between Type IIa and Type IIb events. Thus, classifying an organism based on its appearance can be misleading. One needs a grasp of its genetics, its genotype, if one really wishes to appreciate its relationship with other organisms.

Another related analogy was provided by the great and sadly late microbiologist Carl Woese.[1] Woese proposed that bacteria, that great lump of life that dominates our world, was in fact two living domains. His ideas were thoroughly rejected at the time of their proposition, only for Woese to see his idea vindicated by later DNA analysis.

The opposition to his idea was frequently personal and unpleasant, and Woese suffered academic exclusion for a time. However, by the turn of the millennium the old order had been swept away and it was immensely gratifying to many to see his ideas dominate microbiology.

As well as classification, Woese proposed a fundamental shift in thinking about the inheritance of genes. He saw that gene transfer between species was a significant shaper of their evolution. He suggested, subsequently, a much copied idea that bacteria shared large bodies of genes, particularly in the past. There are some – admittedly tenuous and whimsical – links here with supernovae. One might compare the sharing of genes by microorganisms with the sharing of matter between stars through common envelopes or accretion. The acquisition of new matter may muddy the waters in the phylogenetic tree and therefore in the classification of supernovae.

---

[1]Woese, Carl R.; Nigel Goldenfeld (2009). "How the Microbial World Saved Evolution from the Scylla of Molecular Biology and the Charybdis of the Modern Synthesis". *Microbiology and Molecular Biology Reviews* **73** (1): 14–21.

Indeed, the title of this chapter draws on Woese's work on the sensitive interplay between the advances of molecular biology and its initial conflict with biological phylogeny. In Greek mythology Scylla and Charybdis were two monsters opposing one another across a straight. Sailors wishing to cross this were forced to get too close to one to avoid the other.

In the context of supernovae the conflict is between the potential overflow of information from the automation of supernova capture versus the desire to fit the new events into the old classification framework upon which much of our understanding rests. Woese identified the struggle between the need to accommodate new data from biochemistry, and later DNA technology, with the classification of Archaea and their relationship to other prokaryotic organisms. Microbiological classification, which had primarily been based on phenotype, simply took time to accommodate the large flux of data coming from the (then) new technological approaches.

What Might the New Scheme Be Based Upon?

A new classification system can be built using the same underlying principles as the old, or take a new route. Various options are available that would allow suitable classification of supernovae that would encompass all events. Classification could be built on any of the following:

- Spectra (and hence chemistry) at early and late times – the chemotype.
- Light curves/photometry.
- Mechanics.

The first strategy would be the simplest and merely evolve the current taxonomic system, while requiring a re-naming process that has currently gotten somewhat out-of-hand.

The second option, photometry, is probably the least satisfactory option. The light curve can be affected by a large number of variables, many of which are extrinsic to the supernova itself. Moreover, another fly in the ointment is that many supernovae are simply missed at crucial, early times and the data might, therefore, be incomplete. Moreover, there is an incomplete understanding of the link between the mode of the underlying explosion and the light curve subsequently produced.

The third approach is to look at the mechanism of the explosion. Clearly this approach would discriminate between supernovae of different types, but falters because not everyone agrees on how some supernovae arise. SN 2007bi is a case in point, with many seeing this as the first example of a pair-instability supernova, while others propose that this is merely an energetic core-collapse event.

Idealistically, this latter system is preferable, once there is a clear indication of how supernovae occur. The mechanistic system would also incorporate the current Type .Ia explosions as a separate class related to the novae – shell detonations. At the moment there is a move to separate these from the other events, but the mechanism of subdivision is incomplete, and this is somewhat unsatisfying. Finally, how do we fit thoroughly odd explosions such as SN 2005E and PTF09dav into this system?

A mechanistic system would be analogous to classifying organisms based on their genetic, or DNA, composition – their genotype. This system would be robust and free from contamination by other inputs. Yes, lateral gene flow – the transfer of material from other organisms, or, in our case stars – would muddy the waters. Yet biologists have managed to get around this problem adequately, so why not astronomers?

The reality is that any revised system of classification would undoubtedly be based on the chemotype of the explosion, its underlying chemistry based on the spectroscopy of the event. However, it would be gratifying to see another system grow up in place of the current, chemotypic scheme.

A sensible way forward would be to use some form of cluster analysis. At present supernovae classification uses decision trees: essentially a succession of yes or no answers leading to identification. However, supernovae are not simple creatures. There are a plethora of variables that can and do evolve over time. Therefore, a multivariate analysis of supernovae is now needed. In this approach several variables are considered at once, instead of simply splitting the data into nodes using single variables, such as the abundance of hydrogen or silicon.

In a data universe filled with thousands of examples of supernovae an effective statistical approach is needed to categorize the increasing population of explosions that are pouring out of the automated search programs. From this we will unveil supernovae in all their complexity while identifying the common features that allow their underlying mechanisms to be fully elucidated. Cluster analysis is a familiar tool used by biological phylogenists for decades and has been successfully adapted to classify stars in the HR diagram.[2] The same statistical approaches can be brought to bear on the wealth of observed supernovae.

Simple alternative schemes can be devised that identify the key elements in the spectra of the explosion at early, middle and late times. Many such programs exist to fit model spectra to observed ones. Classification can then be accomplished using data from the spectra at each time interval, perhaps awarding a single letter (or number) code for each principle element at each interval or its relative strength. An additional code could be awarded for the shape of the light curve or the width of spectral features.

The end result will be a three or four letter code for each event, fully categorizing the explosion and into which all earlier supernovae, for which good data exists, can be fitted. Such a scheme can be as simple or as complex as desired, and the output can then be clustered accordingly. Of course, all of this can be automated as well. All the astronomer has to do is point and shoot, then mine the data at the other end. There already exist central repositories for supernova data. A little extra analysis on this data would pull the increasing amounts of disparate information together rather than leave the busy astronomer to mine the literature for similar events that occurred many years previously.

Biologists are all very familiar with this process. For example drosophila geneticists established the comprehensive digital tome Fly-base in the early 1990s to accommodate the vast amounts of genetic and population data that was pouring out of research. Similar work was done for the weed *Arabidopsis thaliana* and for maize – both organisms this author had more than a passing interest in a

---

[2]See http://www.clustan.com/clustering_v_decision_trees.html Clustan analysis.

decade ago. These databases allow very speedy online analysis of experimental data. Similarly most biologists working with RNA and DNA sequences have turned to online data analysis for fast returns on their sequencing endeavors. A section of DNA becomes part of genetic phylogenetic tree within hours of isolation, and very little sweat is broken in doing so.

The establishment of similar systems in astronomy would intensify the rate of data analysis and recovery while simultaneously reducing some of the workload. The computational power has been there for some time, it's just a question of implementing it. The results will be faster classification of supernovae on mechanistic grounds and a greater understanding overall. Moreover, the underlying nature of the relationship each new explosion has with pre-existing events will become clearer, both at a visual and mechanistic level. In the end we will have a clearer understanding of the processes that lead to these intriguing explosions.

Conclusions

These are very exciting times for supernova discovery and in terms of the development of our understanding of the underlying physics of these explosions. Rapid development of computing power and new ideas about how supernovae can be generated is driving rapid change to a field that had seemed to stagnate in the 1990s. Roll on the next decade. There is much more to come.

Glossary

Accretion Disc A disc of material collected from interstellar space or a companion star that spirals onto the surface of another star, thereby adding to its mass

Accretion-induced Collapse The process by which a white dwarf star, most likely made of oxygen, neon and magnesium, undergoes collapse to form a neutron star after accreting material from a companion star. Collapse occurs once the mass exceeds the Chandrasekhar limit (approximately 1.37 solar masses)

AM CVn System A cataclysmic binary consisting of a carbon-oxygen white dwarf and a helium donor star, usually another white dwarf made entirely of helium. During quiet phases helium is accreted onto the CO white dwarf until it becomes hot and dense enough to ignite. This leads to a nuclear explosion, which expands the layer leading to brightening. These novae contain helium but are hydrogen-free. AM CVn systems have short periods of less than 75 min. It is proposed that as these systems evolve, mass transfer rates fall until there is a final flash that may be observable as a Type .Ia supernova

Balmer Lines A type of hydrogen emission line caused by an excited electron dropping back to its ground state from a higher energy level

Black Hole An object at which the escape velocity exceeds the speed of light. Hypothetically, these consist of an enveloping event horizon where the escape velocity matches light speed and a singularity at its center. The singularity is a mathematical construct of infinite density and infinitely small diameter. The existence of the singularity is perhaps questionable

Blue Giant and Supergiant Stars Objects of high luminosity with B- or O-class spectra and diameters several times that of the

Sun. Supergiant stars have masses 8–30 times that of the Sun, while the giants may have comparable or slightly greater mass

Broad-line Type Ic Supernovae These are supernovae that show particularly high expansion velocities (greater than 20,000 km per second) in their ejecta, sometimes referred to as hypernovae. The term refers to the width of emission or absorption lines in the spectra. These are progressively stretched (broadened) as the material giving rise to them moves faster in opposing directions. This is caused by the expansion of the supernova debris. The Type Ic suffix refers to the absence of hydrogen and helium in their spectra. Long gamma ray burst supernovae are broad-lined and are almost exclusively Type Ic

Broad-Wings (Spectral Feature) When energy is emitted or absorbed the spectral line has a width that is influenced by a number of factors. In some supernovae the spectral line is broadened substantially, and this may imply a high gas density where absorption occurs and the influence of gas velocity relative to the observer

Central Engine Supernovae Supernovae driven by jets emerging from some form of central power source, usually a black hole or neutron star. Often referred to as "hypernovae" as a consequence of the higher kinetic energies in the ejecta

Chandrasekhar Limit The mass limit at which the support of a star against gravity by electron degeneracy fails. At the critical mass the velocity of the electrons reach the speed of light, and no further gain in mass is permissible. For a stellar core with a composition dominated by oxygen and carbon this value lies near 1.37 solar masses, but rises to 1.44 solar masses for an object made of iron. The difference is due to the different number of electrons, which increases with atomic number from carbon to iron

Carbon-Nitrogen-Oxygen (CNO) Cycle Technically two interwoven cycles that produce helium from hydrogen using carbon nuclei as a catalyst. Hydrogen adds sequentially to a carbon seed, building it up until an element is created that releases helium and regenerates the original carbon nucleus

Collapsar (Type I and Type II) A model for the formation of a gamma ray burst that involves the formation of a stellar-mass black hole. Type I collapsars are formed from prompt collapse of the core to form a black hole, while Type II form after some delay when material falls back onto the young neutron star

Common-Envelope Phase A stage in the evolution of some binary star systems where the lower mass star becomes consumed by the expanding envelope of the primary star. Both stars then spiral toward one another. In the space of a few thousand years the envelope is dispersed into space, through the conservation of angular momentum, leaving either a merged single star or two closely orbiting stars

Contact Binary A rare binary system where two stars orbit one another so closely that they begin to share their outer envelope. They may then move apart or merge, depending on their respective masses and the amount of angular momentum in the star system

Core-Collapse Supernovae A supernova of a massive star driven by the collapse of its iron core to form either a neutron star or a black hole. A few stars may undergo core-collapse following electron-capture on elements produced by carbon fusion

Decretion Disc One of those terms that seems to have emerged in the last few years but for which there is no precise definition anywhere on the Internet or in a dictionary. The term relates to Be-stars and effectively means a disc of material surrounding the star. Most models for Be-stars indicate the disc has "Keplerian motion," meaning that it is in a stable orbit around the star. This contrasts with the term "excretion disc" proposed by Noam Soker and others for a disc of material shed by a star undergoing a merger and expanding outward. A Be-star is a type of rapidly rotating massive B-class star that is embedded in a disc of material shed from its surface

Deflagration A type of burning typified by combustion in gas engines. Heat transfers through the fuel faster than the burning front advances by simple conduction. The burning front moves at less than the speed of sound within the fuel. This accounts for some Type Ia supernova at all stages and probably the earliest stages of burning in most Type Ia supernova before a detonation occurs

Detonation The process of combustion seen in diesel engines. Combustion is explosive, with a shockwave moving through the fuel faster than heat transfer and ignition by conduction. The speed of transfer exceeds the local speed of sound within the material. The material heats because of compression until it ignites

Electron Degeneracy A special type of very dense material in which the electrons are arranged in restricted energy levels. Electron degeneracy creates a pressure through repulsion that holds some evolved stars up against gravity. Electron degeneracy is restricted by the speed of the component electrons. If the mass exceeds around 1.44 solar masses for iron-rich material, or 1.3682 solar masses for neon, oxygen and magnesium-rich material, the velocity of the electrons approaches light speed. At this point relativistic effects become important and electron degeneracy fails

Erg Unit of energy equivalent to 10^{-7v} joules or 100 n joules (10 millionth of a joule)

Excretion Disc An expanding equatorial disc or torus of material cast off by a star perhaps as a result of a stellar merger or an increase in its rate of rotation or stellar winds

Gas Pressure The pressure created by moving ions, atoms or molecules that is proportional to their temperature and inversely proportional to their volume. Low mass stars are supported in their entirety by gas pressure

Hertz A unit of frequency equivalent to a number of *events* per second. This can be the passing of wave crests or troughs, or the number of rotations of a planet or star per second

Hydrogen-Alpha Emission ($H\alpha$) A specific line of emission of light by an excited hydrogen atom's electron as it decreases in energy levels. Light is in the red portion of the visible spectrum

Hypergiant Stars Very rare stars with luminosities hundreds of thousands to millions of times that of the Sun. Masses vary from

30 to perhaps 300 or more times that of the Sun. Eta Carinae is probably the most famous hypergiant in the Milky Way. Several dozen hypergiants are known across the nearby universe, and a little over two dozen in the Milky Way and Magellanic Clouds

Hypernovae A rather badly misused term referring to very energetic supernovae. Originally coined for supernovae associated with long gamma ray bursts, this reflected a misunderstanding regarding their total energy output. Also applied variously to pair-instability supernovae, hypothetical Population III supernovae or other ultra-luminous explosions. The term can and often is used sensibly and specifically to Type Ib/c supernovae showing jets or other fast-moving collimated outflows

Intermediate Mass Black Hole (IMBH) A theoretical black hole with a mass in the hundreds or thousands of times that of the Sun. These may be produced by particularly massive stars formed through stellar merger or the accretion of large masses of gas onto a smaller stellar mass black hole. Some observational evidence exists for these in young, massive globular clusters in starburst galaxies, such as M82

Joule Unit of energy (work done) equivalent to one Newton of force applied over 1 m of distance

Light Curve A construct of emitted radiation (usually made in the visible spectrum) over time. Light curves reveal the pattern of light emitted by supernovae, allowing both determination of spectral class and an approximation of the total energy emitted by an object during the period of observation

Luminous Blue Variable (LBV) A type of post-main sequence hypergiant star of spectral class O or B that undergoes periodic giant eruptions. These eruptions can rival supernovae in brightness and may shed a substantial amount of mass. Also known as S Doradus variables after the prototype

Magnetar A highly magnetized neutron star (see below) with a surface magnetic field over a trillion times that of Earth. Magnetars may be produced by particularly fast- rotating progenitor stars (or more precisely their cores) or from the most massive stars that fail to form black holes

Main Sequence The region on the HR diagram occupied by stars for 90 % of their lives. Here they remain largely stable, fusing hydrogen to helium

Main Sequence Turn-Off A region on the HR diagram made conspicuous in low and intermediate mass stars where they abruptly leave the main sequence, forming a tail leading to the red giant branch. This is the region where the most massive stars have ceased fusing hydrogen to helium and their inert helium cores are now contracting and heating up

Mass Loss The process by which stars lose mass through stellar winds, eruptions or decretion discs associated with stellar mergers. Rates of mass loss are highly uncertain for massive stars and subject to much controversy

Mergeburst The name given to an outburst of a star caused by its merging with another star, brown dwarf or possibly a planet. The release of orbital energy causes the envelope of the star to heat up and inflate, causing the star to brighten considerably

Molecular Cloud A massive cloud of hydrogen, helium and dust in which much of the component hydrogen atoms have paired up to form diatomic molecules. Most stars are believed to form from the gravitational collapse of these objects

Neutrino A quirky particle produced by the weak nuclear force during nuclear reactions, including some fusion reactions, nuclear decay and Urca processes (see below). These ephemeral particles carry a small amount of mass.

Neutron Degeneracy A type of degeneracy pressure where quarks inside neutrons provide an outward force that can keep neutron stars held up against gravity. The upper mass limit of neutron degeneracy's ability to support a neutron star against gravity lies somewhere between 2.5 and 3 solar masses

Neutron Star A star with roughly 1.2–1.5 times the mass of the Sun produced by the death of a massive star or accretion-induced collapse of a white dwarf star. These are roughly 20 km across and are composed almost entirely of neutrons. Free electrons, iron nuclei and protons may be found in the outer layers of these stars. Many rotate rapidly and emit radiation from confined polar regions. These appear as pulsars, as the beam of radiation sweeps across our line of sight. A fraction of these have particularly intense magnetic fields and rotate slowly. These are known as magnetars. A small number also spin at hundreds of times per second and are known as millisecond pulsars

Nucleosynthesis The process by which new elements are created during nuclear reactions. These can be in stellar cores, in stellar flares, through cosmic ray collisions with other particles or from the Big Bang

Pair-Instability Supernovae Supernovae driven by the loss of radiation pressure within particularly massive stars. At very high temperatures radiation, in the form of gamma rays, is removed, and the core collapses under gravity. The ensuing nuclear fireball converts carbon and oxygen into copious amounts of iron and nickel. The star is completely destroyed in the process that bears a superficial resemblance to a Type Ia supernova (below)

P Cygni Spectrum A spectral feature consisting of a blue-shifted absorption feature and an emission feature that may be somewhat red-shifted. This is produced by an expanding shell of matter moving away from a central object. Commonly seen in supernovae or around some luminous blue variable stars

P Cygni (Star) The prototype of its class; a luminous blue variable that underwent an eruption in 1600, shedding around 0.1 solar masses of material

Photometry The capturing of light from a luminous object at specific wavelengths. These were initially confined to really narrow bands of wavelength, from U (ultraviolet) through various visible bands to I (infrared). Supernovae may be measured in all of these plus others lying in the X-ray or radio portions of the spectrum. Type Ia supernovae are typically brightest in the B (blue) band and also at infrared (I) bands

PP Chain Actually three chains (PP I through PP III) by which stars fuse hydrogen to helium in a string of reactions. Stars with masses of more than twice that of the Sun use an alternative process called the CN or CNO cycle (again, actually, two cycles) to generate energy at a faster rate

P-Process A type of nuclear reaction where protons are added sequentially to seed nuclei. These reactions derive from the CNO cycle on white dwarf stars undergoing novae outbursts and may occur in other settings where there is an abundance of hydrogen fuel mixed with intermediate mass nuclei that can absorb them. The products are proton-rich (or more precisely neutron-deficient) isotopes of elements; many of these isotopes are susceptible to beta-plus (positron) decay

Protostar A large, young star-like object that is still undergoing the processes of formation and is contracting. The radiation emitted is produced from the release of gravitational potential energy, and any nuclear reactions are of limited importance

Pulsational Pair-Instability Supernovae Supernovae that may bang more than once. Pair instability (see above) occurs, but the energy released is insufficient to destroy the star, which merely pulses instead, shedding a large amount of mass. After this unction the star cools, then contracts. If sufficient mass has been shed in the pulse, the core begins stabilize with the fusion of oxygen, and the star then continues its evolution to form an iron core and a standard core-collapse supernova. If not, pulses may continue until sufficient mass has been shed for the stellar core to become stable. Each outburst may emit energy equivalent to a supernova in its own right (around 10^{43} J)

Radiation Pressure A type of pressure produced by the collision and interaction of energetic radiation (mostly gamma rays) with ions and electrons in the core of a star. Radiation pressure has minimal importance is a star like the Sun. However, stars with masses exceeding roughly twice that of the Sun depend on this pressure to hold their cores up against gravity

Red Giant and Supergiant Stars Cool stars of spectral class K or M (and in at least one occasion, class L) that are hundreds or tens of thousands of times brighter than the Sun. The supergiants are larger and more luminous, with radii that would extend out past the orbit of Jupiter were they to replace the Sun, whereas red giants are smaller, with radii extending to perhaps the orbit of Earth or Mars. In general, the radii increase with mass until about 20 solar masses, then decrease again as radiation pressure begins to whittle away at their mass

r-Process The process by which neutrons are rapidly added to seed nuclei of intermediate mass. These processes require very high abundances of neutrons and may occur around the proto-neutron star in a core collapse supernova or in the small, dense accretion disc formed when two neutron stars merge. Elements above lead, with high abundances of neutrons, are produced by this process. Most of these elements are radioactive

Ro-Ap Star A rapidly rotating A-class main sequence star that has a stronger than average magnetic field and may show odd abundances of some elements on its surface

rp-Process A derivative of the p-process where protons are rapidly added to seed nuclei of intermediate mass stars or higher. The process again produces proton-rich (neutron-deficient) isotopes. These reactions may be confined to the surfaces of neutron stars that are accreting hydrogen from a nearby companion star and may be associated with X-ray bursts on these stars

Solar Mass A mass equivalent to that of the Sun, or 1.99×10^{30} kg

s-Process The process by which neutrons are added slowly to elements of intermediate mass, building them up to lead (atomic number 82 and mass number 207) in the Periodic Table. The process is driven by the formation of neutrons through nuclear reactions occurring in asymptotic giant branch (and possibly red supergiant) stars. Neutrons are present at lower densities than found in the r-process (a few billion per cubic centimeter), limiting the rate and scope of these reactions to elements lighter than lead

Spectroscopy The process by which light from an object is split into its component wavelengths by a prism or diffraction grating. The light contains absorption and or/emission lines corresponding to specific elements, which allows the composition of the gas emitting or transmitting the light to be ascertained

Speed vs. Velocity In an utterly reckless fashion these terms are largely interchangeable in this book. However, technically speed is simply how fast an object is going, or more precisely the distance traveled in a unit of time. Velocity is a vector, so not only does it have the same qualities as speed but it has a direction as well

Starburst The process by which a galaxy suddenly (over 10 Ma or less) produces a large number of stars, perhaps numbering in millions. Many of these stars are found in massive star clusters resembling the less showy but still rather beautiful globular clusters seen today around mature galaxies

Stellar Mass Black Hole A black hole with a mass 3–20 times that of the Sun and produced by the collapsing core of a massive star, or through the merger of two neutron stars

Sub-Dwarf These come in two loosely interchangeable forms. Population II main sequence stars (with masses less than that of the Sun) are smaller, hotter and denser than other main sequence stars. These form a strip running parallel to the main sequence, but on the hotter, bluer side. O- and B-class sub-dwarf stars are different. These are the cores of former red giant stars that lie in a portion of the HR diagram called the horizontal branch. The stars in question have shed or have otherwise lost the bulk of their hydrogen-rich envelopes and are burning helium in a shell around their inert carbon core. Outside this they fuse hydrogen in a narrow layer atop the helium layer. They are smaller and considerably hotter than the Sun

Sub-Giant A star with a mass similar to that of the Sun but a few times larger in diameter. These stars have completed core hydrogen fusion, and this fuel now burns in a shell around the contracting helium-rich core

Super-Massive Black Holes Black holes found at the center of all large galaxies, with masses exceeding 100,000 times the mass of the Sun. Their origin is uncertain, but it is intimately linked to the formation of the central bulge of spiral and elliptical galaxies

Supernova Energy The total amount of energy radiated by a supernova can be found (or at least approximated) by calculating the area under its light curve. For the non-mathematical this is the process of integration. A typical supernova radiates between

10^{51} and 10^{52} erg, or 10^{44-45} J. These ridiculously large numbers can be brought a little closer to home by comparing them to typical nuclear explosions on Earth. The biggest nuclear blast (the 58-megaton Tsar bomb) is approximately 10^{27} times less energetic than a run-of-the-mill supernova. Or in other words, a supernova is one followed by 27 zeros bigger – quite powerful, really!

Supernova Imposter An eruption of a star – most likely a luminous blue variable – that is not a supernova but emits an amount of radiation equivalent to a supernova. SN 2008S is a well-known example that caused some dispute in the astronomical community. The eruptions that produce this light are most likely similar or equivalent to the giant eruptions of LBVs and emit up to 10^{43} J of energy, but usually one or two orders of magnitude less energy than this. The cause of these eruptions is unknown

Thermonuclear Supernova A supernova driven by the release of nuclear energy, through the explosive fusion of carbon and oxygen. The most common are Type Ia events, although pair-instability supernovae and the shell-eruption supernovae discussed in Chapter 14 are also thermonuclear in origin

Triple α Process The process, uncovered by Fred Hoyle, through which helium nuclei combine in threes to form carbon-12 nuclei, arguably the most important nuclear reaction in the universe, not only from the perspective of life but from the requirement of this nuclear reaction to produce all the other heavy elements that are found in the universe

Type Ia Supernova A supernova occurring on a white dwarf star that has approached the Chandrasekhar limit and ignited its store of carbon and oxygen fuel. This fuel burns, producing large quantities of iron group elements. Spectroscopically speaking, these are defined by the presence of a prominent silicon absorption line at 635.5 nm and abundant iron

Type .Ia Supernova Pronounced "point one (long) a." A type of thermonuclear supernova involving the explosive burning of helium on the surface of white dwarf star. Proposed by Lars Bildsten and Ken Shen, such eruptions account for a sizable proportion of short Type Ia-like events and may contribute much of the calcium in the universe. Type .Ia events probably originate with the descendants of AM CVn systems

Type Ib Supernova Supernovae that contain helium and prominent intermediate-mass elements. Hydrogen is lacking. Probably exclusively mostly originating from massive stars that have shed their hydrogen envelopes

Type Ibn Supernova Archetypal supernova SN 2006jc showing evidence for excited helium. Produced by the collision of shells of matter deficient in hydrogen but rich in helium. Narrow lines of emission are produced by ultraviolet from the shock breakout illuminating and exciting the surrounding material

Type Ic Supernova Subdivided into broad and narrow line types. Broad lines indicate very high velocities in the ejecta that are often associated with long duration gamma ray bursts that emit material in jets at relativistic speeds. Type Ic supernovae lack evidence for hydrogen and helium in their spectra.

Type IIa Supernova Type Ia (thermonuclear supernovae) that by definition lack hydrogen. These show hydrogen, but this is at later times, implying the collision of the supernova's hydrogen-deficient blast wave with surrounding hydrogen-rich matter. This material is thought to be from a red giant companion star that has previously donated material to the white dwarf

Type IIb Supernova A supernova displaying hydrogen in its spectrum at early times, but these are then "lost" and replaced by prominent lines of helium. Supernova SN 1993J is a well-known example. In comparison with Type IIa events a Type IIb supernova might be reproduced by the collision of the blast wave from a Type Ib supernova with surrounding hydrogen-rich gas

Type II-L Supernova A supernova displaying hydrogen in its spectrum but for which there is no plateau phase. This indicates that the amount of hydrogen in the outer layers was too low to allow for a prolonged phase where hydrogen ions were recombining with free electrons, barring the formation of a flattening of the luminosity in the light curve

Type II-P Supernova A "conventional" hydrogen-rich supernova produced by the death of a red or blue supergiant star following core collapse. The light curve of the supernova contains a flat portion with nearly constant luminosity caused by the cooling and recombination of hydrogen ions and electrons

Type IIn Supernova A Type II supernova that contains a mixture of narrow and intermediate width hydrogen emission lines in its spectrum. The narrow lines are produced by flash ionization of low density hydrogen gas surrounding the star, while the intermediate lines are produced by excitation from collisions of this hydrogen-rich material by the supernova blast wave as it pushes outward. SN 2006tf is a well-known example

Urca Process The process by which electrons are variously absorbed and released by nuclear reactions in the degenerate cores of some stars. These processes release neutrinos and their antimatter equivalents, causing the cooling of the stellar interior. Urca processes are important mechanisms by which neutron stars, white dwarfs and the cores of some giant stars cool following the cessation of nuclear fusion reactions

Wolf-Rayet Star A very luminous star that has shed its hydrogen and sometimes helium outer layers through stellar winds or through transfer of material to a nearby companion star. Such stars display varying amounts of helium and carbon and/or oxygen. Some transitional stars contain small amounts of hydrogen and nitrogen instead, and these are believed to be O-class stars that are losing mass on their way to form carbon or oxygen-rich Wolf-Rayet stars

Index

A
Accretion, 43–45, 126, 129, 140, 141, 159, 162, 168, 169, 172, 189, 193, 226, 257, 258, 260, 263, 266, 267, 281–283, 285, 294, 308, 316, 328, 346, 355
Accretion disc, 44, 64, 66, 128, 141, 159, 160, 165, 190, 263, 266, 282, 295, 351, 359
AGB. *See* Asymptotic giant branch (AGB)
AG Carinae, 92
AM Canum Venaticorum (AM CVn), 293–299, 317, 351, 362
AM CVn. *See* AM Canum Venaticorum (AM CVn)
AM CVn binary, 294, 295
AM CVn variable, 294
Angular momentum, 25–28, 42, 64, 68, 70, 133, 134, 141, 145, 157, 230, 247, 261, 328, 329, 332, 333, 353
Asymptotic giant branch (AGB), 95, 176, 177, 182, 197, 263–265, 287, 311, 326, 330, 359

B
Balmer lines, 37, 351
BATSE, 119
Beppo-SAX, 120–123, 130, 136, 137, 139, 142, 143, 152
BeX binary, 189
Binary system, 17, 154, 155, 157, 158, 166, 168, 173, 256, 258, 259, 267, 282, 289, 293, 295, 323, 328, 339, 353
Black hole, 27, 30, 43, 48, 66, 83, 102, 117, 126–129, 131, 133–136, 139–142, 145, 151, 154–164, 167, 168, 170–173, 189, 193, 203, 230, 236, 246, 247, 250, 300, 306, 345, 351–353, 355, 360

Broad line type Ic supernova, 352
B-Type Star, 52, 87, 189

C
Capella, 13, 14
Carbon-12, 17, 56, 181, 361
Carbon-13, 17, 288
Carbon-fusion, 57, 177, 178, 183–185, 195, 353
Cassiopeia A, 67, 69, 163
Chemotype, 18, 33, 87, 105, 225, 347, 348
Chromium-48, 296, 298
Circumstellar medium (CSM), 100, 107, 111, 115, 134, 148, 283
CN(O) cycle, 24, 54, 55, 169, 196, 288, 293, 352, 357
Cobalt-56, 76, 100, 116, 212, 221, 223, 298, 300
Collapsar (model), 124, 126–129, 131, 134, 135, 142, 151, 152, 159, 248
Conservation of momentum, 26, 28, 42, 68, 261, 353
Convection, 20–24, 29–31, 47, 52, 54, 55, 63, 69, 86, 87, 93, 104, 167, 180–186, 194, 231, 287, 288
Crab Nebula, 175
Cryonuclear fusion, 268
CSM. *See* Circumstellar medium (CSM)
Cygnus X-1, 66, 154–160, 162

D
Deflagration, 317
Detonation, 36, 92, 105–107, 111, 114, 119, 161, 172, 191, 196, 212, 227, 249, 257, 263, 264, 268–271, 273, 278, 280, 285, 291, 293, 296, 297, 300–305, 309–319, 338, 348, 354

366 Index

Deuterium, 16, 45
Double-degenerate, 259, 265–267, 282, 283, 293, 320

E

Electron, 4–7, 16, 17, 19, 21, 22, 37, 42, 49, 50, 52, 59, 61, 63, 67, 68, 75, 87, 94, 99, 102, 108, 115, 122, 125, 135, 140, 148, 162, 177, 183, 185–188, 200, 201, 212, 259, 294, 311, 351, 352, 354, 356, 358, 362, 363
Electron-capture, 162, 300, 306, 335, 353
Electron-capture supernova, 102, 163, 175–198
Eta carinae, 88, 92–94, 96, 97, 105, 107, 109, 111, 200, 203, 208, 214, 305, 339–342, 355

F

Fall-back, 43, 65–67, 73, 102, 129, 139, 142, 145, 153–164, 189
First dredge-up (-out), 181, 182
Forward shock, 73
Fusion, 14–16, 18, 24, 25, 47, 54–60, 67, 87, 88, 93, 177, 178, 180–185, 187, 190, 194–197, 201, 202, 218, 229, 260, 266, 268, 287, 288, 293, 294, 296, 302, 315, 326, 353, 356, 358, 360, 361, 363

G

Gamma ray burst (GRB)
  GRB 010921, 162
  GRB 011211, 136, 137, 140
  GRB 020124, 139
  GRB 020813, 140, 141, 143
  GRB 030329, 136
  GRB 031202, 136
  GRB 060218, 136, 236
  GRB 060614, 156, 158, 159, 162
  GRB 081007, 136
  GRB 970228, 121, 123, 125
  GRB 970508, 121–125
  GRB 970514, 148
  GRB 980326, 136
  GRB 980425, 66, 130–132, 134, 136, 137, 140
GCD. *See* Gravitationally confined detonation (GCD)

Giant star, 11, 13, 32, 51, 90, 166, 167, 191, 194, 195, 227, 266, 327, 331–334, 360, 363
Gravitationally confined detonation (GCD), 269, 271, 273, 280
GRB. *See* Gamma ray burst (GRB)
GRO J1655-40, 155

H

HE0107-5240 (star), 42
He I emission, 29
Helium, 6, 41, 87, 127, 161, 166, 176, 201, 240, 257, 278, 285, 325, 351
Helium fusion (Triple-α process), 55, 56
Helium shell, 70, 146, 187, 218, 260, 268, 287, 300, 312–314, 317, 318, 326
H_2 emission, 105, 351, 363
Henize 2-10, 171, 172
Hertzsprung-Russell diagram, 3, 10, 11, 24, 32, 89, 182
Hot bottom burning, 194, 288
Humphrey-Davidson limit, 89–91, 93, 96
Hypergiant, 12, 97, 193, 212, 215, 354, 355

I

IMBH. *See* Intermediate mass black hole (IMBH)
Initial mass function, 42
Intermediate mass black hole (IMBH), 170, 173, 355
Intermediate mass star, 30, 43, 49–51, 57, 95, 176–178, 180–182, 191, 195, 197, 259, 264, 356

J

Jeans mass, 41

L

LBV. *See* Luminous blue variable (LBV)
Luminous blue variable (LBV), 27, 88–90, 92–95, 97, 102–107, 109, 111, 113, 115, 117, 118, 150, 152, 169, 172, 192, 193, 199, 200, 206, 209, 305, 315, 323–325, 335–338, 355, 357, 361
Lyman-α line, 6

M

Magnetar, 28, 69, 70, 116, 142, 162, 229–237, 244, 246–249, 355, 357
Main sequence, 11–15, 20, 23–26, 28–30, 45–54, 83–118, 153, 166, 170, 171, 177, 180–183, 190, 191, 196, 200, 217, 218, 265, 273, 293, 317, 325–328, 337, 345, 356, 359, 360
Mergeburst, 323, 327, 328, 330, 333, 335, 338–342, 356
MGG-9 (M82), 171
MGG-11 (M82), 171, 227
M85 OT20061, 335
M31 RV, 338
M-type star, 187

N

Neutrino, 16, 17, 59, 61–63, 69, 77, 107, 126, 153, 155, 159, 184–186, 195, 201, 231, 282, 356, 363
Neutron star, 27, 28, 30, 48, 62–65, 67–70, 77, 83, 86, 117, 119, 120, 128, 129, 139–142, 145, 153, 155, 157, 159, 163, 164, 166–168, 171, 184, 186–189, 191, 193, 195, 203, 229–231, 236, 246, 258, 266, 300, 345, 351–353, 355, 356, 359, 360, 363
NGC 300 OT, 336–338
Nickel-56, 60, 76, 99, 109, 112, 116, 129, 133, 139, 145–147, 149, 156, 158, 159, 162, 188, 195, 197, 207, 209, 210, 212, 219–224, 234, 243, 255–258, 261, 268, 275, 277–281, 285, 289, 291, 292, 296, 298, 300–302, 316, 318, 335
Nitrogen-14, 17, 87, 181

O

Optical transient (OT), 336–338
Orion Nebula, 6, 29
OT. *See* Optical transient (OT)
O-type star, 6
Oxygen fusion, 59

P

Pair instability (PI), 26, 27, 83, 94, 106, 111, 112, 116–118, 132, 172, 197, 199, 200, 202–211, 217–228, 230, 235, 243–245, 248, 249, 280, 298, 319, 320, 341, 348, 355, 357, 358, 361
P-Cygni, 37–39, 92, 98, 99, 108, 113, 209, 242, 331
Photodisintegration, 61
Photometry, 3, 8–9, 110, 163, 347, 357
PI. *See* Pair instability (PI)
Population I star, 30
Population II star, 26, 43, 129, 197, 227, 360
Population III star, 26, 27, 129
Positron, 16, 17, 59, 76, 94, 201, 205, 294, 296, 298, 318, 358
PPI. *See* Pulsational pair instability (PPI)
Proton-proton chain, 16–17
Protostar, 29, 44–46, 64, 165, 358
PSR 1257+12 (millisecond pulsar), 141
PTF 10aaxi, 149
PTF 09uj, 100
Pulsational pair instability (PPI), 111, 117, 118, 172, 200, 204–211, 217, 221, 227, 235, 243–246, 248–250, 341, 358
Pyconuclear fusion,

R

R136, 166, 168, 169
R136a, 250
R Corona Borealis stars, 286, 287, 311
Reverse shock, 73, 155, 245
Rho Cassiopeia, 89, 91, 92, 97, 193
rp-process, 167, 168

S

SAGB. *See* Super asymptotic giant branch (SAGB)
Sakurai's object, 260, 287
S Doradus variables. *See* LBVs
Single-degenerate, 259–263, 265, 267, 273, 282, 293
SNR 0509-67.5, 265, 320
Spectroscopy, 3, 13, 14, 164, 213, 223, 348, 359
Spectrum
  absorption, 4, 5, 7
  continuous, 4
  emission, 5, 7
Spinstars, 27, 226
Super asymptotic giant branch (SAGB), 102, 162, 163, 179, 191, 194, 195, 306
Super-Chandrasekhar supernova, 275–284

Supergiant, 12, 13, 46, 55, 65, 70, 72, 77, 90, 91, 101, 105, 141, 144, 149, 150, 154, 168, 192, 195, 212, 213, 230, 247, 286, 288, 323–325, 328, 330, 337, 341, 342, 351, 352, 358, 359, 363
Supernovae
  PS1-10awr, 248
  PS1-10ky, 248
  PTF 09atu, 241
  PTF 09cnd, 241
  PTF 09cwl (SN 2009jh), 241
  PTF 10cwr (SN 2010gx), 241, 242
  PTF 09dav, 310–313
  SCP 06F6, 110, 241, 242
  SN 1000+0216, 223, 224
  SN 2213-1745, 223, 224
  SN 1987A, 73
  SN 2006aj, 136, 233, 235–237
  SN 2008am, 208
  SN 2002ap, 147, 222
  SN 2005ap, 8, 216
  SN 1996aq, 138
  SN 2009bb, 145
  SN 2005bf, 232–235, 237
  SN 2007bi, 199, 217–227, 248, 280, 344, 345, 348
  SN 2002bj, 297, 298, 300, 301
  SN 1999br, 162
  SN 1998bw, 37, 66, 76, 130–134, 136, 139, 147, 149, 222, 233
  SN 1979C, 240, 247
  SN 2000ch, 97
  SN 2005cs, 102, 189, 313
  SN 2002cx, 161, 305–310
  SN 1997cy, 131, 147, 149, 264
  SN 1997D, 162–164
  SN 2008D, 72
  SN 2001dc, 162
  SN 2009dc, 280, 281
  SN 2003dh, 136, 147
  SN 2005E, 102, 313–318, 348
  SN 2008es, 211–217, 239, 240, 247
  SN 2011fe, 271–273
  SN 2003fg (SNLS-03D3bb), 275, 278–281, 291, 293
  SN 2005gl, 106
  SN 2003gm, 97
  SN 2006gy, 108, 199, 206
  SN 2006gz, 280, 281
  SN 2008ha, 160–162, 305–310
  SN 2005hk, 302–310
  SN 2008hw, 136
  SN 2002ic, 263, 264, 274
  SN 2007if, 280, 281
  SN 2008iy, 110, 111
  SN 1993J, 75, 362
  SN 2006jc, 362
  SN 2003jd, 138, 222
  SN 2010jp, 149–152
  SN 2001ke, 137
  SN 2005ke, 263
  SN 2002kg, 97, 103
  SNLS-04D2dc, 72
  SN 2003lw, 136
  SN 2010mc (PTF 10tel), 104
  SN 2009md, 193–195
  SN 1994N, 162, 189
  SN 1996r, 100
  SN 2008S, 361
  SN 1991T, 263, 264, 276, 278, 280, 291, 307
  SN 2006tf, 108–115, 117, 204–209, 213, 214, 219–222, 230, 243–246, 344, 363
  SN 1994W, 102, 109, 112
  SN 2010X, 301
Supernova imposter, 96, 103, 111, 112, 175, 176, 192, 305, 337, 361
Supernova remnant, 28, 37, 64, 67, 73, 74, 100, 142, 154, 162, 229, 232, 269
Supernova shockwave(s), 3, 71, 72, 99, 107, 109, 113, 114, 204, 209, 229, 233, 248, 262, 264, 273
Supernova types (classes)
  Type 1.5, 196–197, 264
  Type Ia, 295–305, 308
  Type Ib, 36, 70, 145, 315, 362
  Type Ibc, 72, 142–149
  Type Ic, 70, 71, 103, 131, 138, 144, 147, 154, 160, 211, 240, 242, 245, 249, 352
  Type II-L, 75, 211–217, 240, 247, 362
  Type IIn, 97–99, 102, 104, 106, 109, 112, 149, 162, 175, 191, 199–228, 230, 264, 363
  Type II-P, 99, 107, 110, 195, 363
Super-soft x-ray source, 260, 261
Swift, 72, 143, 145, 157, 158, 243, 287, 334

T

Taxonomy (of supernovae), 33–34
Theta Orions C, 29
Thorne-Zytkov object (TZO), 166, 168, 173

Titanium-44, 76, 100, 298, 307, 315, 318
Trapezium star cluster, 29
TZO. *See* Thorne-Zytkov object (TZO)

U
Urca process, 184–186, 356, 363

V
V838 Monceratua, 324–330
V4332 Sagittarii, 330–331
V1309 Scorpii, 329, 331, 334, 340

W
WC star, 88, 103
White dwarf
    carbon-oxygen, 197, 255–257, 267, 269, 271, 275, 276, 278, 285, 287, 288, 294, 299, 300, 316, 318
    helium, 287, 294, 295, 308, 316
    oxygen-neon-magnesium, 57, 293, 300
WN star, 169
Woese, Carl, 346
Wolf-Rayet (WR) star, 27, 53, 65, 87, 88, 117, 127, 128, 133, 135, 139, 140, 150, 152, 159, 161, 169, 230, 244, 363
WO star, 88

X
X-ray flash (XRF)
    XRF 060218, 143
    XRF 080109, 146

Z
Zeeman effect, 28

Lightning Source UK Ltd.
Milton Keynes UK
UKOW06f1247040516

273538UK00004B/20/P